TOWARD SUCCESS IN BIOMASS CONVERSION TO AFFORDABLE CLEAN ENERGY

TOWARD SUCCESS IN BIOMASS CONVERSION TO AFFORDABLE CLEAN ENERGY

The Story of KiOR and the Merits and Perils of
Developing Economically and Environmentally Sustainable Biofuels
to Chase Down Global Warming and Limit Destructive Climate Change

DENNIS N. STAMIRES *and*
STEPHEN K. RITTER

ARCHWAY PUBLISHING

Archway Publishing books may be ordered through booksellers or by contacting:

Archway Publishing
1663 Liberty Drive
Bloomington, IN 47403
www.archwaypublishing.com
844-669-3957

Interior Image Credit:
Figures courtesy of Dennis Stamires, with exception of Figure 9A reprinted with permission from Ind Eng Chem Res 2008;47(3):742-747.

ISBN: 978-1-6657-4320-4 (sc)
ISBN: 978-1-6657-4321-1 (hc)
ISBN: 978-1-6657-4322-8 (e)

Library of Congress Control Number: 2023908128

Print information available on the last page.

Archway Publishing rev. date: 07/19/2023

CONTENTS

1.0 INTRODUCTION

The oil crisis in the 1970s had us all driving a little scared. Long lines waiting to fill up your car with gasoline seemed anti-utopian, especially for a country like the U.S., given the American love affair with the automobile and the necessity of vehicles of all types for transportation and commerce. The crisis revealed in very real terms our growing dependence on energy production and supply. However, that period was just a stitch in time, and we endured.

Now, 50-plus years on, we look back at those days and they do not seem so remarkable. We have since experienced significant global fluctuations in the availability and cost of energy, in particular transportation fuels, for a number of reasons: supply, demand, geopolitical tensions, and—most important—access to economically and environmentally viable technology.

The 1970s oil crisis did achieve one thing: It helped rekindle interest in using dedicated or waste biomass, directly or indirectly, as a fuel or to produce transportation fuels, resulting in a number of public and private research initiatives to develop technoeconomical solutions. Some of these efforts have come to fruition in the form of bioethanol derived from corn, sugarcane, and other crop sugars and biodiesel derived from methanol or ethanol combined with soybean oil, other vegetable oil, or animal fat.

Yet, the efforts of chemists and chemical engineers in enabling a biobased industry to compete against fossil fuels and Big Oil have been limited so far. The infrastructure for obtaining raw materials such as crude oil and natural gas, processing it in established high-capacity refineries, and distributing it quickly in pipelines and by other high-volume means is too well-established. In fact, in the past decade we dipped into a fear-factor lull because petroleum and natural gas production found new life, driven by advances in technology, including hydraulic fracturing, or "fracking," defying the odds that fossil fuels will run out. In the U.S., domestic crude oil and natural gas production actually increased enough so that in 2020 the U.S. became the surprising global leader and a net exporter of fossil fuels, a far cry from the 1970s. The current availability and, in reality, modest cost of energy have kept thoughts of practical biobased production of fuels on the back burner, though ongoing interest fluctuates with energy prices and other variables.

Still, the fact looms that crude oil is eventually going to become scarce enough or too costly and/or environmentally unfriendly to procure and process, making biomass conversion technoeconomically favorable. Another mitigating factor concerns whether we should be burning our fossil fuels at all to produce energy—and unwanted carbon dioxide, methane, and other emissions—or whether we should be using these resources to make commodity chemicals instead [1]. At some point, bioenergy will be the most attractive option, if not for energy security and affordability then to meet the demands of a world population that is consuming more energy per capita. One variable that will influence the use of biobased fuels is the expectation that by 2050 solar-energy/solar fuels technology and electric-vehicle technology will be advanced to the point for substantial or even complete replacement of combustion-engine vehicles. Even so, transportation fuels will still be needed for mass transportation in airplanes and delivery of goods by long-haul trucking, trains, and ships. Biobased fuels could meet those reduced needs without using fossil fuels. We should not discount either that environmental stewardship stemming from human impacts on pollution, greenhouse gases, and global warming with destructive climate change, associated with extracting raw materials, processing them, and producing and consuming energy, will one day rank at the top of global society's to-do list.

Where does that leave us? No matter what the future holds with regard to globally sustainable, affordable, clean, and low or net zero or negative carbon energy, it is imperative that we keep in mind the availability and cost lessons learned thus far about bioenergy production and decarbonization. We need to seriously consider where we stand at present with economically feasible and environmentally responsible scalable-to-commercial technologies without government subsidies or tax incentives, and act accordingly.

1.1 HISTORICAL PERSPECTIVE

In this account, we focus primarily on the U.S., as it is the most significant consumer of energy in the world, for now. The biomass-to-fuels conversion that the 1970s oil crisis started driving us to was not a new idea, even back then. Humanity had relied on wood and other biomass for millennia, to produce energy and feed animals that provided transportation. During the Industrial Revolution, virtually all raw materials were derived from renewable natural plant or animal sources or natural mineral resources in seemingly inexhaustible supply. Scientists and engineers over the past century and longer have tinkered with how to use cellulosic materials and plant sugars to synthesize fuels and chemicals; making ethanol and producing naval stores (turpentine, rosin, tar, and pitch derived from pine forests) go back quite a bit longer.

The concept of a biobased industry nearly caught on about 90 years ago. In May 1935, approximately 300 industrial, agricultural, and scientific leaders met in Dearborn, Michigan, as part of the of the Chemurgy movement—that is, the promotion of chemical and industrial use of organic raw materials. One outcome was the "Declaration of Dependence upon the Soil and the Right of Self-Maintenance." The goal was to express the "inalienable right" to explore a new frontier for the industrial utilization of agricultural crops [2].

Yet, as new technologies developed, we turned to coal and

eventually started refining petroleum. Chemurgy lost out to petro-chemicals during and just after the Great Depression and World War II, although the concept of energy independence has remained active in political circles. Then 50 years ago Earth Day was celebrated for the first time, and spurred by the oil crisis, brought renewed public attention to human-driven environmental degradation and finite fossil-fuel supplies. But by the 1970s, petroleum refining was well-developed, and the challenges of mass production of liquid fuels from biomass at an affordable cost and requiring no new infrastructure or new type of automobile engine just never worked out. Oil was too plentiful, easier to process, cheaper, and ingrained into the fabric of society.

In 1971, U.S. President Richard M. Nixon's Administration attempted regulatory control of fuel prices to counter a quickly increasing inflation rate. However, overall energy availability and fuel costs became much worse, in part because of the subsequent 1973 oil embargo by the Organization of the Petroleum Exporting Countries (OPEC). The embargo stemmed from U.S. support of Israel during the Yom Kippur War, which at the time was the latest in a series of oil-rich Middle East conflicts. A shortage and rationing of transportation fuels back then led to uncertainty—part of the aforementioned scary feeling. For example, some gas stations in Southern California allowed customers to purchase only 10 gallons of gasoline per week, and schools organized carpools for transporting students to save on fuel. Some gas stations were forced to close, with hastily produced handmade signs "Sorry ... No Gas" becoming a common sight (Figure 1).

As a world economic recession deepened, coupled with the energy crisis, the Nixon Administration decided in late 1973 to establish "Project Independence." This effort was aimed at reducing, by 1980, U.S. reliance on imported oil. Ideally, this project would have achieved a steady balance of domestic supply and consumption of crude oil and transportation fuels—that is, achieving national energy self-sufficiency. Furthermore, with growing environmental awareness, the hope was that the project would help reduce smog, acid rain, and greenhouse-gas emissions through a commitment to energy conservation and to developing alternative, next-generation clean-energy sources to avoid a global humanitarian crisis as the world's population increased.

A part of this effort was the Corporate Average Fuel Economy (CAFE) standard, enacted in 1975, with a goal to prompt development

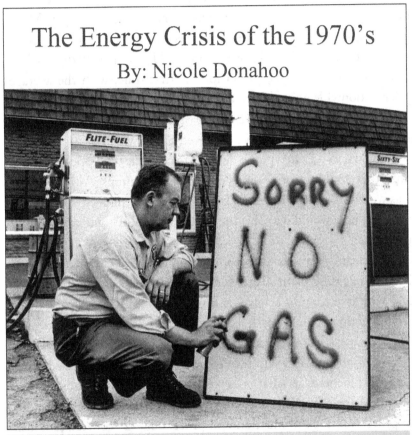

Figure 1. An example of the spreading energy crisis of the early 1970s is illustrated in this photo of the title page of a report written by Nicole Donahoo, a concerned elementary school student at Harbor Day School in Newport Beach, California, (and granddaughter of Dennis Stamires a coauthor of this book), who organized a carpool with her classmates to save gasoline.

of more efficient engines to improve gas mileage in cars and light trucks. In addition, this effort came about in part to help stretch the gasoline supply and sustain corn growers by replacing some petroleum-derived automobile fuel with ethanol; ethanol had been considered as an automobile fuel from the time the first cars were designed, but it had taken a backseat to gasoline.

Project Independence failed though, and U.S. dependence on oil imports actually grew by 1980. The scary feelings endured. Many gas pumps in those days were designed for a maximum fuel price of 99.9 cents per gallon. However, in the U.S. in late 1979, gasoline prices topped $1.00 per gallon for the first time, with the average price nationwide going above $1.00 by the end of 1980. This created confusion: If you pumped $1.00 worth of gas, as shown on the register, you might actually have delivered $2.00 worth of gas, as some station owners adjusted pump delivery to coincide with price increases until they could retrofit pumps or install new pumps. Although that was worrisome at the time, the bigger question remained—no matter the cost, how would we solve the future availability and affordability problems?

It's clear by now that this conundrum is as recalcitrant as biomass itself in releasing plant sugars for chemical processing to make fuels and will take time to resolve, and not without bigger steps, namely federal action to help spark advances in technology. In this regard, "The Biomass Research & Development Act of 2000" was enacted as bipartisan U.S. legislation to spur development of biobased products and bioenergy as a national priority [3]. The law introduced four technical areas for R&D activities: (1) develop crops and systems that improve feedstock production and processing, (2) convert cellulosic biomass into intermediates that can be used to produce biobased fuels and products, (3) develop technologies that yield a wide range of biobased products that increase the feasibility of fuel production in a biorefinery, and (4) analyze biomass technologies for their impact on sustainability and environmental quality, security, and rural economic development. The legislation was further amended by the

Food, Conservation & Energy Act of 2008 [4] and reauthorized in the Agriculture Improvement Act of 2018 [5].

These congressional actions encouraged efforts over more than a decade for fledgling companies to pursue commercialization of biomass to fuels and chemicals technologies in the name of energy independence. Yet, none of these efforts so far have managed to supplant petroleum as the main source of fuels and chemicals. Modest successes have been achieved, but anything close to a U.S.-biobased industry is lacking. Efforts elsewhere in the world have met with varying degrees of success on a national or regional level, but any complete, economically sustainable, and environmentally acceptable global biobased industry is still far away.

When the day comes that we find ourselves relying predominantly on biomass for transportation fuels, assuming such a day does come, we will need a viable commercial-scale technology. The technology will need to operate using low-grade, inexpensive, nonfood biomass feedstocks including recyclable and waste materials; take advantage of existing infrastructure; and be environmentally friendly. While use of ethanol blended with gasoline together with biodiesel and the development of hybrid electric vehicles has started to bridge this transition, none of those options will likely be a complete solution. The traditional mix of gasoline, diesel fuel, and jet fuel is really needed—but sourced entirely from virgin and waste biomass. On that front, a number of biobased technologies have been developed and start-up companies created in recent decades to get a foothold in the fuels and chemicals marketplace. These companies have struggled, however, largely because of technoeconomical shortcomings.

One good case study is that of KiOR. This Houston-based company created in 2007 employed biomass catalytic cracking (BCC) technology used by Dutch company BIOeCON and was subsequently supported financially for further development in large part by Vinod Khosla and his capital investment firm Khosla Ventures. BIOeCON was founded in 2005 by Paul O'Connor, Dennis Stamires (a coauthor of this book), and Armand Rosheuvel. O'Connor, a chemical engineer, became KiOR's Chief Technology Officer. Stamires, a heterogeneous catalysis and solid-state physics and chemistry expert, served as Senior Science Advisor to the KiOR management team, a member of KiOR's Science Advisory Board, and supported the R&D technical team. O'Connor and Stamires received shares of KiOR stock when the company was formed. Rosheuvel was an investor and finance director at BIOeCON, but he was not involved with KiOR.

Beginning in 2007, until declaring bankruptcy in November 2014, KiOR spent close to $1 billion to prove that a single-reactor in situ thermocatalytic conversion process using conventional petroleum-refining catalysts to turn biomass into transportation fuels is not scalable to commercial-size plants, and in fact is not economically feasible using current technology. The reasons behind KiOR's failure run deeper than technology problems. Critical managerial mistakes happened along the way, reflective of a start-up company getting off on the wrong foot, having too much money available from investors looking for a big win, and simply being reckless—a modus operandi doomed to fail from the outset.

WHAT'S IN A NAME?

The biofuel company KiOR's name has an uncertain meaning. One line of thought is that it was proposed by leading investor Vinod Khosla because it sounds similar to d'or, to be "golden" or "excellent," or possibly because it is reminiscent of Dior, the fashion designer. Other possible meanings are "the chosen one" or "one of many talents." But there is no clear etymology. Since about 2018, Kior is increasingly being used as a person's name, at least in the U.S., according to Social Security Administration statistics, possibly meaning "origin" or "popularity." It also is the name of a line of electronic personal care devices.

What KiOR did gain is the experience of going through the whole sequence of conventional developmental gate stages. The company's staff constructed and used gram-scale lab-testing equipment, kilograms-per-hour pilot plants, a 10 ton-per-day semiworks/demonstration unit, and finally the world's first 500 ton-per-day commercial biomass-to-fuels plant. At peak capacity, this commercial plant was projected to produce as much as 92 gallons of bio-oil per dry ton of biomass, resulting in 13 million gallons of fuels per year. This progression sheds useful light on the stage of scale-up at which researchers may reliably expect to recognize process limitations, with the benefit of any red flags along the way to signal a warning for the need to alter course by making adjustments to the process or the need to take a more radical course and replace the technology altogether.

KiOR's story is easy to follow, because the company's history is outlined in the patent literature and has been well-documented by the press, academic studies of business practices, and internal company communications among legal papers in court cases. The patents, press reports, and other published materials remain a rich source from which useful technical information can be extracted and used by scientists and engineers. For example, in 2016 *Biofuels Digest* published an illuminating description of KiOR's history in a series of articles by Jim Lane [6], and in 2019 Joseph Mohorčich extensively discussed KiOR in a Sentience Institute white paper on successes and failures in biofuel commercialization as a proxy for scaling up other types of biotechnology, such as cultured meat [7].

KiOR's commercial plant in Columbus, Mississippi, known as Columbus I (Figure 2), did produce close to 1 million gallons of transportation fuels in 2013, mostly gasoline, which was a notable achievement. The company sold it to oil refineries for blending with petroleum-derived fuels. Besides being less than 10% of the projected production volume, KiOR's return on investment was quite low and prospects for improving were low, leading to the company's demise.

Those production numbers by further comparison were a drop in the bucket, for example, relative to total U.S. annual transportation fuels production of more than 200 billion gallons. This volume equals about 30% of total U.S. annual energy production, according to the U.S. Energy Information Administration [8]. In addition, because of the sheer volume, the U.S. currently leads the world in biofuels production, making 18.2 billion gallons of the 40.1 billion barrels produced globally in 2019, according to the Energy Information Administration and the International Energy Administration [8,9]. The U.S. accounts for about 45.5% of total global biofuels production, followed by Brazil (26.5%), Germany (2.9%), Argentina (2.7%), and China (2.6%). In a sign of progress, the total amount produced globally has increased more than fivefold over the past 20 years.

We report 2019 values, given that the COVID-19 pandemic introduced an unexpected yet additional significant variable affecting global fluctuation in the availability and cost of energy: U.S. energy production dropped 5% during 2020 and petroleum consumption in the U.S. was at its lowest level since the 1970s; 2021-2022 data indicate that energy production is returning to pre-pandemic levels. Perhaps the most interesting statistic is that in 2020 renewable energy production in the U.S. surpassed coal for the first time since 1885. U.S. energy production should continue growth throughout 2023 and beyond, but another question to raise is when might renewable energy production surpass petroleum consumption?

A quick note on the size of the problem: Global energy demand is forecast to increase about 27% by 2040, according to the International Energy Agency's World Energy Outlook, taking into consideration an anticipated world population growth from 7.9 billion in 2022 to 9.9 billion in 2050, with most of the growth in Asia. Along the way, environmental protection technology advances will be required to produce affordable, clean, renewable fuels. An interesting discussion of this expected global energy demand problem and the associated resource challenges is provided in a report, "Seeing the World from a Different View," which was presented by retired U.S. Navy Vice

Admiral Philip H. Cullom at the Advanced Bioeconomy Leadership Conference on April 4, 2019 [10].

The analysis provided here comes at an important time, as economic experts had been forming a consensus that we are heading toward another fuel crisis, at least before COVID-19. One harbinger of this potential global economic crash is the growing demand and insufficient supply of low-sulfur (0.5 weight %, or less) diesel fuel used for ships, trains, agricultural machinery, and long-haul trucks, as well as fuel oil for heating [11-12]. The demand is in response to tighter atmospheric emissions standards mandated by the International Maritime Organization, which became effective globally on Jan. 1, 2020. Oil refineries already draw on limited available supplies of light-sweet crude, which is used primarily to produce low-sulfur fuels. The increased demand was projected to potentially bump up the price of the crude to $160 to $200 per barrel. We will have to wait and see whether this prediction ever comes to pass, so far it is unclear, as shipping slowed globally during the pandemic and starting in late 2022 the potential for a global recession was looming. In addition to clamping down on sulfur, in December 2022 the U.S. Environmental Protection Agency took a further step to reduce pollution from diesel-powered trucks, delivery vans, and buses by lowering allowable emissions of nitrogen dioxide, with the biggest impact on the use of these vehicles while they are idling or moving at low speed; U.S. EPA plans additional rules on cutting emissions from these vehicles in 2023. One can envision that biofuels could provide a partial solution by increasing the supply of low and ultralow sulfur and nitrogen diesel that could be blended with noncompliant petroleum-derived diesel to produce a product meeting the new clean-air standards.

The technology to make this needed biodiesel is already available. For example, KiOR developed a process for commercial-scale plants that could handle up to 2,000 tons of dry nonfood biomass per day. This process could produce low-sulfur and low-nitrogen biodiesel

Figure 2. KiOR's Columbus, Mississippi, commercial plant under construction.

with yields of more than 80 gallons per ton of dry biomass at a production cost close to $3.00 per gallon, based on $70 to $80 per ton of biomass feedstock. However, the company's management decided not to test this new technology on commercial scale at Columbus I.

Ongoing fluctuations in energy supply, demand, and cost, as we will return to later on, justify continued research and capital investment in biomass fuel production by federal funding agencies and private investors. The benefits include creating attractive business opportunities and jobs, on top of the security benefits granted to an evolving global society that is increasingly dependent on a stable supply of sustainable, affordable, and low, net zero, or negative carbon clean energy. Our global society is further in dire need of implementing fail-safe measures for controlling ground, water, and atmospheric pollution along with greenhouse-gas emissions to protect Earth's environment and reduce the impacts of global warming and destructive climate change.

With all these elements in mind, this account describes KiOR's story. It is offered as a guide to help find the best technoeconomically sustainable pathways forward for further development of scalable-to-commercial biobased commodity clean fuel and chemical technologies.

1.2 TECHNOLOGY SOLUTIONS

Fuel technology advances primarily come from oil companies, and therefore they provide a starting point for developing large-scale biofuels production. For example, one of the pioneers of the field, Theodore C. Frankiewicz, working at Occidental Petroleum, obtained a patent in 1981 outlining the first use of a zeolite-based catalyst to convert crude bio-oil obtained from pyrolyzing waste biomass to light hydrocarbons to make chemicals and fuels [13]. A number of others were reporting similar advances at about the same time [14-16].

Frankiewicz used a two-stage processing technology involving a circulating fluidized bed (CFB) flash thermolytic reactor to convert waste plastic and biomass materials into crude bio-oil, followed by thermocatalytic upgrading of the bio-oil to light deoxygenated hydrocarbons in a second CFB reactor. The patent described a catalyst that incorporated a synthetic small-pore pentasil silicalite-type zeolite; natural and synthetic zeolites are microporous aluminosilicate crystalline materials. The catalyst had been patented in 1975 by Mobil Oil Co. and coded as Zeolite Socony Mobil–5, or MFI ZSM-5, which is commonly referred to as plain ZSM-5. While the active catalyst component is the crystalline zeolite, it is typically compounded in different proportions with extenders such as kaolin clay in water slurries containing binding agents and then spray-dried to form catalyst microspheres that are calcined by heat treatment. These catalyst microsphere particles are used commonly in fluid catalytic cracking (FCC) units in petroleum refineries around the globe to produce transportation fuels as well as commodity and specialty chemicals.

The valuable performance advantage of the Frankiewicz two-step process stems from thermolytic devolatilization of the biomass in the first reactor, using sand as the heat-transfer medium, and separating out the metal-containing ash and char generated before the crude bio-oil produced is introduced into the second reactor and brought into contact with the cracking catalyst. This separation step improves catalyst performance and prolongs catalyst lifetime on-stream compared with a one-reactor in situ system in which those unwanted by-products are not removed and thus are allowed to come in contact with the catalyst with ill effects.

However, because of insufficient published pilot-plant data and still limited knowledge of catalytic biomass thermolysis, and especially because of unavailable information on the physico-chemical properties of ZSM-5 zeolites at the time, Frankiewicz in the early 1980s did not recognize the detrimental effects on the catalyst caused by the acidic hydrothermal conditions in the reactor and in the catalyst regenerator unit, which operates at a higher temperature. Although Frankiewicz was removing the biomass ash, metals, and char from the first reactor, as well as noncondensable gases such as carbon monoxide, carbon dioxide, and methane, he neglected to remove the acidic water generated during the biomass conversion before transferring the crude bio-oil to the second reactor for catalytic upgrading. The hot acidic water/vapor phase degrades some zeolite active sites via protonation of crystal lattice oxygen atoms, which forms hydroxyl groups that upon calcination/dihydroxylation produce water and cause lattice reconstitution. The result is loss of some zeolite crystallinity, which affects catalyst stability, activity/longevity, and selectivity, even more so as the catalyst recycles through the hot regenerator over time. We will return later to the effects of bio-oil acidity, the importance of removing indigenous biomass metals to prevent catalyst degradation, and oxygen content issues in upgrading crude bio-oil.

Frankiewicz made an additional contribution though by introducing the new concept that the extender and binder components, which help disperse the zeolite and shape and strengthen the catalyst microspheres, must each have a low surface area of about 25 m²/g. This property is similar to the low surface area of pure, refractory alumina or silica, which are metal oxides used as catalyst support materials. Controlling the surface area is a useful means of reducing formation of coke (a carbon-rich solid) that can clog zeolite pores and deactivate the catalyst, adversely affecting product yields. Another key aspect of Frankiewicz's advances involves going beyond reducing the surface area of the catalyst microspheres to also reduce the pore volume of the particles, which also lessens coke formation.

Taken together, these attributes lead to denser particles, and by becoming much denser, the particles' volumetric heat capacity increases. The bulk material thereby becomes a more efficient heat-transfer medium, a fundamental property required for quickly transferring a large amount of heat from the regenerator to the reactor during catalyst recycling to achieve efficient thermolytic devolatilization of the biomass feed in the mixing zone of a single-reactor system. The topics of heat transfer and thermocatalytic performance also are discussed in detail further on.

However, when Frankiewicz's second-stage process was tested in a CFB pilot plant, it proved to be inefficient. Because it operated with long residence times and high temperatures, the process produced excessive amounts of gases, resulting in low liquid hydrocarbon product yields.

Further developments using CFB reactors for biomass ultrafast thermolysis in the 1980s did prove to be much more efficient when mixing reactants with catalysts and/or heat-transfer media to provide controllable short contact times and moderate temperatures for producing liquid hydrocarbons. For example, this was originally demonstrated by the pioneering work of D. S. Scott and coworkers at the University of Waterloo [17-19]. Further still, pioneering research by Iacovos A. Vasalos at the Chemical Process Engineering Research Institute (CPERI) and Aristotelian University of Thessaloniki, in Greece, led to a patented process, assigned to Amoco (Standard Oil), where he previously worked, for retorting hydrocarbon-containing materials such as oil shale, coal, and tar sands in a CFB reactor to produce liquid hydrocarbons suitable for upgrading to transportation fuels [20].

Another notable advancement by Vasalos and coworkers was the modification of a conventional oil refinery FCC unit for use as a CFB reactor to devolatilize waste biomass, using circulating hot sand as a heat-transfer medium [21]. Vasalos, working with Angelos A. Lappas and their colleagues at CPERI, used the modified FCC system with a catalyst containing the small-pore ZSM-5 zeolite in ultrafast thermocatalytic conversion of biomass to bio-oil and light hydrocarbons containing relatively low amounts of oxygen [22]. These hydrocarbons, however, still contained an appreciable number of oxygenated molecules coming from the biomass cellulose and carbohydrates, usually 10 to 25 weight %. This molecular mix requires subsequent hydrodeoxygenation and/or hydrocracking processing to obtain lighter and fully deoxygenated hydrocarbons suitable for use in transportation fuels and in some specialty chemicals.

As an aside, straight thermolytic conversion of biomass to bio-oil leads to material with high oxygen content, up to 40%, roughly the same level as in the biomass itself. This level of oxygen, besides being too high to be directly refined within the existing infrastructure of today's oil refineries, causes the bio-oil to become viscous and unstable as the oxygenated hydrocarbons polymerize while in storage and to become difficult to handle—lowering its heating value.

This is also a good time to mention that significant amounts of hydrogen gas are required for removing the oxygen and upgrading the crude bio-oil via hydrocracking and hydrotreating to produce light hydrocarbons as the useful end products of biomass conversion. Hydrogen production and consumption is one of the major process cost variables in refineries, which will be discussed in due course.

In addition to those efforts, start-up companies sponsored by private investors appeared just after the 1973 oil embargo to develop technologies for producing upgradable hydrocarbons from petroleum-based waste materials. For example, Deco Industries, a joint venture of actor John Wayne, Greek shipping magnate George P. Livanos, R. William Chambers, and others, produced transportation fuels from shredded used automobile tires and municipal wastepaper and plastics at a facility in Orange County, California. Stamires served as a technical consultant to Deco in developing its process; Deco stands for Duke Engineering Company, derived from Wayne's nickname "Duke." This waste-to-energy initiative involved a continuous-feed auger thermolysis reactor operating under reduced pressure in the absence of air, with the option of varying the temperature along the length of the reactor. This system enabled the hydrocarbons formed to depart quickly from the reaction zone before undergoing secondary reactions leading to undesired products. A segment of the overall technology has been patented [23].

Whereas Deco's main objective was to help alleviate the urgent need for petroleum-derived fuels, the company was also interested in providing an environmentally acceptable means for disposing of used rubber tires—another consequence of meeting our transportation needs—along with waste plastics that degrade slowly and leach pollutants in landfills. Deco's approach not only had potential to meet both those objectives, but also to reduce the need to expand landfills, when available land in Southern California was at a premium, and further to reduce methane, carbon dioxide, nitrogen, and sulfur compound emissions (hydrogen sulfide and mercaptans) produced from landfills that contributed to air pollution and smog, leading to atmospheric ozone depletion at a time when global warming and destructive climate change were gaining greater attention.

For example, starting in the early 1960s on the Palos Verdes Peninsula near Los Angeles, mature landfills leaking these gases were creating an unhealthy environment for the highly populated local communities. This problem was mostly alleviated by drilling wells into the landfills and inserting perforated pipes to collect the gases, which subsequently were processed in a facility using a pressure swing absorption (PSA) process with molecular sieve/synthetic zeolite sorbents to separate the methane, other light hydrocarbons, and mercaptans for delivery to local natural gas supply companies for consumer use. Stamires participated in the molecular sieve sorbent selection and modification and the optimization of the PSA process, which was successful in curbing the need for expanding landfills, reducing landfill greenhouse gas and other emissions to the atmosphere, and at the same time producing a usable fuel. These and related technologies have been used commercially in Japan and Korea since then, and different attempts have been made in the U.S. [24].

Of additional historical interest, it should also be mentioned that Livanos, as part of his environmental protection advocacy, engaged Deco to develop new equipment for cleaning up ship bilge water. Separating oil and lubricants from water mixtures generated by engines, generators, pumps, and other mechanical devices used in large cargo ships, oil tankers, cruise ships, and ferries permits the purified water to be dumped as ships travel and the collected hydrocarbons to be reused or recycled, to the benefit especially of port areas. Livanos installed the equipment in his own shipping fleet managed by Ceres Hellenic Enterprises, a practice that was followed by other major shipping owners. This environmental advancement to clean up ship waste was announced at the Posidonia International Shipping Exhibition, held in 1972 at Zappeion Palace in Athens, Greece, with Livanos, Wayne, Chambers, and Stamires among those participating in the Deco public announcement (Figure 3). Through his close friendship with Wayne, Livanos arranged that Wayne inform U.S. President Nixon, a personal friend, about the Deco equipment and the campaign to clean and protect navigational waters. Nixon in turn passed the technology interest to the U.S. Department of Defense to arrange for Navy ships to install environmental protection bilge oil separation pumps.

Figure 3. Announcement of Ceres Hellenic Enterprises' Separation & Recovery Systems (CERES-SRS) in Athens in 1972 included (rear, left to right) Joseph Defranco (Deco President), Livanos, Stamires, Wayne, and Chambers along with other Deco staff and executives from Sinopec and Nippon Oil Co.

The new measures fit well with regulations from the U.S. Environmental Protection Agency, which had been formed just a couple of years earlier in December 1970 during the Nixon Administration, with encouragement from Wayne, to save land, water, and air from industrial- and consumer-caused pollution and greenhouse gas emissions. Stamires and Wayne also travelled to Japan and China to meet with Nippon Oil and Sinopec/China Petroleum & Chemicals officials to develop cooperative R&D projects for producing clean fuels from organic solid waste and producing chemicals while reducing process gas emissions to the atmosphere.

History has provided a number of these technology stories that influence energy and transportation fuels and Earth's environment. Another is the life and times of supersonic transport (SST) aircraft. Developed in the 1960s at the same time the Apollo program was sending astronuats to the Moon, a fleet of these high-flying airplanes held out promise for faster commercial travel. There were two problems: Noise and damage caused by sonic booms as the planes travel faster than the speed of sound, and the cost for increased fuel use and air pollution from burning the fuel in the stratosphere (~60,000 feet). A main threat from the engine emissions is that nitrogen and sulfur oxides can destroy ozone that is crucial for protecting Earth from the Sun's harmful ultraviolet radiation.

Some solutions were to avoid flying over land especially in heavily populated areas, to fly at lower altitude, and using low-sulfur and low-nitrogen fuels. As debates continued among legislators, federal agencies, environmental groups, aircraft makers, and airline companies, it was clear more research was needed. McDonnell Aircraft and Douglas Aircraft Company, which merged in 1967 to form McDonnell Douglas, had an interest in supplying planes to government agencies and commercial airlines, and as a condition for receiving government contracts the company was required to create an R&D facility. During this period, Chemistry Nobel Laureate Willard F. Libby at the University of California, Los Angeles, known for developing radiocarbon dating, was on the Douglas Aircraft Board of Directors and was appointed to oversee the design and operation of the new research facility. Named Douglas Advanced Research Laboratory (DARL) and located in Huntington Beach, California, the new facility was similar to other corporate R&D labs in those days, such as Bell Laboratories and General Electric Research Laboratory, and brought in leading scientists and engineers such as Howard Reiss at UCLA and James Mercereau and Richard Feynman at Caltech as consultants.

Libby was also a member of the California Air Resources Board and had his own research projects investigating combustion engine exhaust and emissions control. To that end, he expressed an interest in supporting a DARL project, directed by Stamires, working on investigating the interaction of ozone with SST aircraft exhaust gases to assess ozone destruction. In the lab, ozone and exhaust gases were mixed and studied at the concentration, temperature, pressure, and ultraviolet light conditions found in the stratosphere (Figure 4). Gas from the reaction chamber was fed into an electron spin resonance (ESR) spectrometer to measure the amount of free-radical species formed and determine the associated loss of ozone. Preliminary results confirming ozone depletion were submitted in a report to the U.S. Department of Transportation for use in its Climatic Impact Assessment Program.

Libby's influence extended to one of the students he mentored, F. Sherwood Rowland, and his colleague at the University of California, Irvine, Mario Molina, who sorted out the chemistry of ozone depletion. This research resulted in a ban on chlorofluorocarbons (CFCs) used in consumer and commercial applications, with Rowland and Molina later sharing the 1995 Nobel Prize in Chemisry with Paul Crutzen, who studied the effects of nitrogen oxides on atmospheric ozone.

Altogether, these three independent research efforts in the 1960s and 1970s, those of Libby and his group at UCLA, Rowland and Molina at UC Irvine, and Stamires and associates at DARL, all played a role in confirming each project's findings and contributed significantly to developing automobile catalytic converters, phasing out CFCs, and effectively pulling the plug on SST aircraft. The DARL study in particular led to a decision to cut federal funding for SST aircraft in 1971, and also influenced a decision to prohibit SST aircraft from flying over land in the U.S. and decisions from several coastal states to restrict SST aircraft flights.

These side stories provide an example of the interconnectedness of science and subsequent applied technology and how innovation often drives itself in new directions to find solutions to intractable problems, or preventing problems from being created in the first place. Another good example that is now playing out is how global society can solve its future energy demand needs in an affordable and environmentally acceptable manner, as we outline in this account.

Figure 4. Assembled by Stamires and colleagues at DARL in the mid-1960s, this room full of equipment was used to study the effects of jet engine exhaust on atmospheric ozone, which in part led to discontinued U.S. funding for developing supersonic transport aircraft. An ozone generator on the bottom right fed the gas into a reaction chamber above where jet engine exhaust gases were added under conditions found in the stratosphere, with the reaction products analyzed in an in-line ESR spectrometer shown on the left. Similar experiments were carried out using methane and different chlorofluorocarbons to study their impact on atmospheric ozone.

1.3 TERMINOLOGY

Readers of this chronicle by now may have noticed the use of the terms "devolatilization," "thermolysis," "hydrothermolysis," "thermolytic," and "thermocatalytic" in place of the more familiar terms "pyrolysis," hydropyrolysis," and "pyrocatalytic." The reason is that pyrolysis as presently used does not correctly represent the physicochemical conditions that are actually involved in such biomass conversion processes, that is, high temperature in the absence of oxygen.

Historically, the term pyrolysis originates from the ancient Greek words "pyra" and "lysis," meaning fire and disintegration/dissolution. The words "pyr" or "pyra" have a generic and in some cases philosophical meaning, for example, denoting death, funeral, destruction, extinction, and anger. In modern Greek, there are slang expressions and moral stories. For example, when a mother found out that her son lied to her about attending school and instead had gone to a park and played with friends, she was struck with "pir/pyr" and "mania," that is she saw red or became "fire-crazy."

Regarding the terminology of thermochemical processes, perhaps more confusing than pyrolysis are the occasionally used terms "thermal pyrolysis," "thermolytic pyrolysis," or "hydrothermal pyrolysis," which literally mean hot fire, or hot burning, which is nonsensical. Pyros means "of the fire," and fire needs oxygen to exist. As a combustion process, pyrolysis produces gases and ash, not an organic liquid. Thus, the term pyrolysis represents the physical process wherein biomass or any organic matter in the presence of oxygen/air is combusted, producing gases along with smoke and ash; it includes fire and lightning.

A fireworks display, for example, is made possible by using pyrotechnics, a correct use of "pyro," which is also appropriate as noted already for pyromania and for pyromaniacs. A naturally demonstrated pyro effect involves forest fires, as seen proliferating in the western U.S., Australia, Western Europe, and elsewhere stemming in part from global warming and climate change. In this case we see the flames burning trees, other vegetation, and homes, with the smoke plume generated forming cumulus clouds as it rises and cools, which correctly are called pyrocumulus clouds, fire clouds, or flammagenitus clouds, and so on. Volcanoes can cause the same effect.

However, to accurately represent the process of converting biomass into bio-oil by heating in the absence of air/oxygen/flames, the term thermolysis or hydrothermolysis, meaning hot hydroseparation and hydrodissolution, are more correct; devolatilization and hydrodevolitilization, that is the conversion and/or the removal of volatile substances from a solid in the presence of hydrogen, are useful alternatives.

PYROLYSIS, OR NOT?

The term pyrolysis represents the physical process wherein biomass or any organic matter in the presence of oxygen/air is combusted, producing gases along with smoke and ash. It is often used mistakenly to describe thermal biomass conversion to a crude oil, which is more correctly thermolysis, a process of heating biomass in the absence of air/oxygen/flames. Pyrolysis no, thermolysis yes.

Indeed, by the 1990s, knowledgeable researchers in the field began using the term thermolysis in place of pyrolysis in their patents and articles. The pioneering bio-oil producer Ensyn labeled its core biomass conversion pathway "Rapid Thermal Processing," or RTP. And a number of other pioneers and experts working on biomass-to-biofuels conversion processes started using it and devolatilization [25-27].

To understand how this misnomenclature came about requires a brief visit to the 1920s and the prevailing technical thinking at that time related to generating biofuels, primarily from shale, and how the "Godfather of Pyrolysis," William A. Hamor, came up with this name.

An excerpt from a 1922 article in the journal *Chemical & Metallurgical Engineering* explains [28],

> *Another term which is not only cumbersome and unwieldy but in general quite distasteful is that of destructive distillation, a term which creates with the layman an undesirable impression, as was shown by a rural, bewhiskered 'old timer' who, after listening to a paper read at one of our shale conventions, arose and said, 'If you chemists and engineers would quit your "destructive" distillation and do*

a little "constructive" distillation, we would soon begin to get somewhere.' Dr. W. A. Hamor of the Mellon Institute, however, has very splendidly substituted 'pyrolysis' for this term, and the new word is rapidly being adopted and already becoming extensively used by our chemists and engineers."

1.4 MOTIVATION FOR CONTINUOUS IMPROVEMENT

Moving on from these historical notes, after the 1970s petroleum availability increased and lower fuel prices returned for consumers. These achievements impeded national interests in biofuels development and waylaid venture capital investors hoping to profit off R&D projects to form start-up companies to produce next-generation fuels. However, the on-again, off-again cycle of interest and funding repeated. Oil prices in the 1990s sparked development of new technologies and restarted dormant ones for producing transportation fuels from nonfood and waste biomass sources. Growing concerns with greenhouse-gas emissions also began to have more sway in biofuels development.

Bioenergy takes advantage of recycling carbon in the global ecosystem, as opposed to drilling and mining and subsequent use of fossil fuels that adds carbon back to the ecosystem after it had been removed in eons past and stored away underground. Any way one wants to think about whether global warming is real or not, the physics of Earth's atmosphere dictates that, all other variables being equal, processing and burning fossil fuels taken together with all human activities that add carbon dioxide, methane, and other greenhouse gases to the atmosphere means the planet is going to become warmer. And a change in temperature means the dynamics that drive the weather are going to change, sometimes in destructive ways.

Petroleum engineer Eugene Houdry had originally led

development of FCC processes utilizing CFB reactors in the late 1920s, and these systems were being widely used commercially after World War II for converting petroleum crude oil to gasoline, diesel fuel, and jet fuel [29]. However, most biomass conversions using similar technologies today are still being evaluated in pilot plants, demonstration plants, or semiworks plants. No technology has yet been scaled to a commercial-size plant that has consistently demonstrated environmentally acceptable economic feasibility without government subsidies or incentives.

In fact, several start-up companies from recent decades working on next-generation biofuels have changed their operations to produce high-value, biobased specialty chemicals for niche markets rather than fuels, or they have simply gone out of business. The failure to commercialize biomass-to-fuels approaches at scale has occurred for a variety of reasons, including lack of operating funds or unfeasible/unscalable technology, resulting in financial losses to investors and taxpayers. From these financial losses, our premise is that it would be useful to salvage some knowledge regarding why certain technologies are feasible and could work and others do not, to avoid repeating the same costly mistakes.

2.0 KIOR'S STORY: A CASE STUDY AND REARVIEW ANALYSIS OF A $1 BILLION GAMBIT

KiOR's experience is a good example of a lesson learned, because KiOR did it all. The essence of the company's BCC technology involved pretreating raw biomass/carbonaceous material with a catalytic material, such as potassium carbonate or bicarbonate, and then introducing the pretreated biomass into an FCC-type CFB reactor together with a catalyst for thermocatalytic conversion to crude bio-oil suitable for refinery upgrading to transportation fuels and specialty chemicals (Figure 5).

The active component in the compounded catalyst particles was the synthetic double hydroxide lattice layered basic anionic clay, named hydrotalcite (HTC). This material contains magnesium and aluminum hydroxides, with a general formula $Mg_6Al_2CO_3(OH)_{16} \cdot 4H_2O$, described by Stamires and colleagues in a U.S. patent "Mg-Al Anionic Layered Synthetic Clay Having 3R2 Stacking" and in a *Journal of Materials Chemistry* research article "Synthesis of the 3R2 Polytype of a Hydrotalcite-like Mineral" [30]. Besides its use in industrial chemistry, HTC is commonly used as an antacid and as a nanocarrier for drug delivery such as antimicrobial therapy for cancer patients.

At first, without having its own laboratory equipment, KiOR contracted CPERI's Vasalos and Lappas to use their modified FCC pilot plant in Thessaloniki, Greece, to convert biomass to bio-oil

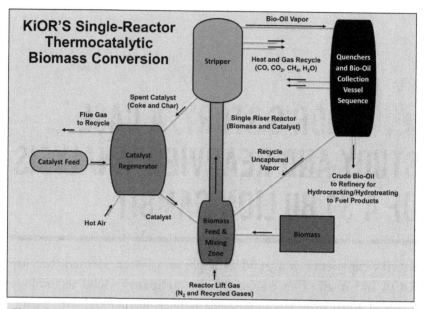

Figure 5. KiOR's Single-Reactor Thermocatalytic Biomass Conversion Process. KiOR employed the same basic process design for gram-scale lab-testing equipment, kilograms-per-hour pilot plants, a 10 ton-per-day semiworks/demonstration unit, and a 500 ton-per-day commercial biomass-to-fuels plant. In this system, a traditional FCC-type catalyst is added to the pretreated biomass feed and the mixture passed through a single vertical riser with proprietary variable geometry capable of providing a residence time for the solid particles of 20-50 seconds and bio-oil vapor residence time of less than 1.5 seconds. A stripper separates entrained vapors with steam from the spent catalyst (containing coke and char) with the crude bio-oil further processed to prepare it for upgrading in typical refinery processes. Some components such as cyclones to separate vapor from solids, vapor and process gas filters, heat-collection vessels, and heat-exchangers are not shown. The single-reactor process produced uneconomical low yields of highly oxygenated bio-oil because biomass conversion by-products (acidic water vapor) led to high catalyst deactivation and the need for frequent catalyst replacement.

using HTC as the BCC catalyst. The company also contracted catalysis expert Avelino Corma at the Institute of Chemical Technology (ITQ) at Polytechnic University of Valencia, in Spain, who had previously been under a consulting contract with BIOeCON, to use ITQ's benchtop-scale reactors for thermocatalytic conversion of biomass to liquid hydrocarbons.

At first, without having its own laboratory equipment, KiOR contracted CPERI's Vasalos and Lappas to use their modified FCC pilot plant in Thessaloniki, Greece, to convert biomass to bio-oil using HTC as the BCC catalyst. The company also contracted catalysis expert Avelino Corma at the Institute of Chemical Technology (ITQ) at Polytechnic University of Valencia, in Spain, who had previously been under a consulting contract with BIOeCON, to use ITQ's benchtop-scale reactors for thermocatalytic conversion of biomass to liquid hydrocarbons.

The test results from CPERI, reported to KiOR in August 2008, were disappointing. In general, bio-oil yields from the pilot plant were low because the catalyst being employed produced large amounts of coke, char, and undesirable gases. The ITQ team delivered its final results to KiOR in January 2009. These results were only interesting from an academic perspective, however, to see how different chemical pretreatments of the biomass feed affected process variables, such as useful product distribution, because like the CPERI results the bio-oil yields were low with high levels of undesirable by-products. Additionally, the catalyst used was quite expensive, as will be discussed later, plus it deactivates quickly and needs frequent regeneration and replacement with fresh catalyst. These early results indicated that the low bio-oil yields and using an expensive catalyst could not support an economically sustainable bioenergy business.

But by this time, KiOR had its own R&D lab in place in Houston. The R&D team was made up of scientists and engineers with substantial experience in chemical and petroleum refining processes, specifically in thermocatalytically converting petroleum to gasoline, diesel fuel, and jet fuel. KiOR's team, led by O'Connor as Chief Technology Officer, included Technology Director Jacques De Deken, Senior Manager of Process & Catalyst Development Robert Bartek, and chemical engineers Peter N. Loezos, Steve Yanik, and Conrad Zhang; later on they were joined by Michael Brady, Charlie Zhang, and Agnes Dydak. Stamires worked as a consultant in a supporting role to the technical team and reported directly to KiOR President and CEO Fred Cannon, a former executive at AkzoNobel and Albemarle.

The first red-flag warning signal about the failing BCC technology was raised by De Deken in the summer of 2008. At that time De Deken was the most knowledgeable technologist in the KiOR organization, as he was receiving and evaluating all the biomass devolatilization test results from the company and from the two European laboratories. In a nutshell, the results led De Deken to conclude the BCC technology was not working and by far was not meeting KiOR's business objectives and should be immediately replaced. De Deken and his team presented the details to O'Connor and Cannon in August 2008. However, O'Connor, in agreement with Cannon, rejected the test results, analysis, and conclusion without explanation—despite the raised red flag—and decided to press on without implementing R&D changes. This outcome created a contentious situation that led to De Deken's departure from KiOR on August 11 after only five months on the job; other key technical personnel were reassigned to different functions, and still others left KiOR at that time.

In his resignation letter, which later was made public, De Deken stated,

> "The strategy in rushing toward demonstrating the BCC technology at a multi-barrel-per-day scale without actual and reproducible corroborating experimental data, under the pretense of self-deception of creating value, is a recipe for technical failure."

In a follow-up letter to O'Connor, Cannon, and rest of the management team, De Deken criticized KiOR's administration,

> "What is even more worrisome is that genuine efforts to establish a dialog about relevant technical issues have been met with systematic attempts to downplay or dismiss virtually every issue as soon as it is brought up. Clearly,

the creation of lasting value is not possible without also developing credible, sound, and robust technology. KiOR's obvious lack of commitment to build a strong and much needed R&D effort to make this possible is a further indication that KiOR is not really serious about developing successful technology."

In retrospect, De Deken's observations about KiOR and the company's future turned out to be correct and came in time, just it wasn't heeded.

Subsequently, Bartek, a chemical engineer with extensive R&D experience in petroleum refining processes, catalyst development, and FCC pilot-plant testing and evaluation, assumed management of the pilot plants and demonstration unit operations, as well as the catalyst development work.

By December 2008, Bartek had accumulated more pilot-plant results confirming the unfeasibility of the BCC technology. On December 7, Bartek sent an email to R&D staff members, copied to Stamires, with a subject line "More Math on BCC."

"I agree we are in a period of denial. We must forget our original conceptions of BCC and must do something radically different to save the project."

In February 2009, Stamires wrote to a team comprised of Bartek, Yanic, Loezos, and Brady to propose that they immediately use KiOR pilot-plant test runs to duplicate published test data obtained from similar pilot plants, including those of competitors using the same biomass feed and with sand as the heat-transfer medium. Stamires was suggesting they work in stealth mode to pursue the calibration and baselining project that had been repeatedly ruled out by O'Connor. This was an urgent matter to resolve, to avoid constructing the new 10 ton per day demonstration reactor with the wrong design—the design of this demo unit was being debated by the management team in the absence of process engineering experts, including Bartek. This bit of recklessness cost KiOR time and money, as the reactors of a pilot plant and the demonstration unit subsequently were built with wrong designs, and later had to be modified.

Continuing this saga, on March 28, 2009, Bartek responded to Stamires' request, with a copy to stealth team member Brady,

> "You had already been hounding me to get sand in the unit. At that time the three of us started on this, I had already accepted the fact that I was a 'dead-man walking' in [Cannon's] organization, and my time at KiOR would be short. So why not one final act of defiance? If you are going to be let go, let's do it for a noble project reason rather than politics. Maybe we could rescue this thing and snatch victory out of defeat we [are] heading into."

Ultimately, Bartek conducted the duplicate test at KiOR's pilot plant, using sand obtained from CPERI for consistency, and confirmed significantly higher bio-oil yields as obtained at CPERI's pilot plant. Bartek and Stamires then proceeded to arrange for Vasalos to visit KiOR's office to work with Cannon to develop a licensing agreement allowing KiOR to use CPERI's pilot plant design in KIOR's pilot plants, demonstration unit, and in the Columbus commercial plant. Subsequently, the meeting with Vasalos and Cannon took place in KiOR's Houston office, and an agreement was signed.

Bartek obtained more catalyst test results from KiOR's pilot plant and from CPERI's pilot plant, including results from round-robin collaborative testing supervised by Vasalos and Lappas. The results obtained from both labs were the same, showing low bio-oil yields with high coke formation and undesirable noncondensable gases, reconfirming previous results of the failing BCC technology. Meanwhile, KiOR was running on fumes financially.

Also at this time, Cannon underwent heart surgery and was out of the office for some time, leading to more managerial bad news and confusing most of KiOR's employees over who was in charge. On March 19, O'Connor notified the staff that he had taken on Cannon's responsibilities "to assume a smooth continuation of our business." Yet, the Board of Directors being uninformed of this managerial change had not commented. In addition, Stamires, who was in contact with Cannon while he was at home recovering, was told by Cannon about an email from O'Connor criticizing him on his leadership and for creating a poor working environment and low morale—overall forcing key personnel to leave the company. To that effect, O'Connor offered Cannon to help solve the managerial problems, and save the company, by taking over some of the CEO's responsibilities. Cannon had no authority to make changes in KiOR's management hierarchy, as that was the role of the Board of Directors.

Further to this management saga, Andre Ditsch, Vice President of Strategy, reacted negatively to O'Connor's actions to assume the role of interim CEO, and criticized O'Connor's performance. Ditsch also claimed for himself some of Cannon's managerial responsibilities during his recuperation and ended up competing with O'Connor for KiOR's leadership. This infighting led to a kind of corporate functional paralysis, preventing KiOR from making sensible company decisions to move forward. For example, there were times when both O'Connor and Ditsch were calling their own separate staff meetings and occasionally pursuing competing management agendas. This situation, which was demoralizing the staff and hindering progress, was finally resolved by the Board of Directors, led by Vinod Khosla. Cannon returned to his position, Ditsch continued in his strategy role, and O'Connor was forced to step down from his position of CTO, though he remained on KiOR's Board of Directors.

In the meantime, Bartek decided to go around the management and take the matter directly to Khosla. On June 3, 2009, Bartek presented his findings and conclusions to Khosla, as the lead investor, in a meeting at Khosla's office in San Francisco. Khosla had taken an unusual approach with KiOR in that his firm was more involved in company operations than a typical investment firm might be. Bartek notified Khosla that the bio-oil yield in the pilot plants was in the mid-40s gallons per dry ton of biomass and that the BCC process would not be able to achieve the 92-gallon performance goal. Khosla requested the KiOR management to find a way to double the bio-oil yield during the next six to eight months. Bartek conveyed this request to KiOR staff members in an email a few weeks later. On June 26 he stated,

> "... *re-emphasized that the BCC process was not working and will be impossible to achieve the bio-oil yield of two barrels per ton of dry biomass, necessary to meet KiOR's business objectives.*"

Bartek, rather than being credited for helping save KiOR, was later demoted and moved away from overseeing the pilot-plant operations and catalyst development work. KiOR continued its established plan and further did not react to published technical and economic sustainability information on competing technologies, hurting the R&D effort. For this "beating a dead horse" mentality, the company paid a high price by wasting time and financial resources, because later KiOR still had to revamp the catalyst and reactors. Bartek resigned from KiOR on January 4, 2011. The departures of De Deken and Bartek fatally hurt KiOR and left the remaining technical personnel disappointed and with serious concerns about the company's future—and their own future in the company—with a hint of failure already in the air.

2.1 DUE DILIGENCE: TECHNICAL ASPECTS OF COMMERCIAL PROCESS DEVELOPMENT

Estimates of plant construction and production costs of transportation fuels at commercial scale, based on experimental data obtained at the early R&D stages, are part of a company's in-depth due diligence for new technology projects to confirm proof-of-concept. These analyses are necessary for predicting the possible success of a risky business, preventing costly mistakes, and calculating needed investment funding. Upon scaling up to commercial-size plants, dominant factors that drive the overall performance of a process become more apparent, that is process limitations and inherent deficiencies, rendering any back-of-the-envelope, pie-in-the-sky, expected margins moot.

Some companies neglect to conduct these conventional gate-stage analyses, or to do so thoroughly, but KiOR did not fall into this category. KiOR's failing bench-scale, pilot-scale, and semiworks unit results were largely correct; however, their interpretation, or rather the company leadership's handling of the numbers, were incorrect. The details of these events will be borne out later, but in short, KiOR publicly presented an unrealistic positive technoeconomic analysis that its process and catalytic materials would be successful and cost-effective on a commercial scale along with a business model forecast predicting sustainable profitability by 2015. The reality for KiOR was that its BCC technology was not going to work at scale—the available data consistently provided a disproof-of-concept.

To get a handle on the technological circumstances in hindsight, the article "A Survey of Catalysts for Aromatics from Fast Pyrolysis of Biomass" by Shantanu Kelkar and colleagues, for example, is helpful [31]. This article describes conversion of poplar wood to bio-oil in a microscale reactor using 0.5 g of biomass in a packed bed located between two beds containing a ZSM-5 catalyst. The researchers tested samples containing zeolites with different silica-to-alumina molar ratios (SARs) and found that the lowest SAR zeolite samples provided the highest yields of aromatic hydrocarbons and the lowest levels of undesired coke. They based this conclusion only on the nascent, short-lived, measured fresh activity and selectivity of the catalyst. However, the study does not, by far, represent the complete performance of the catalyst on-stream in a commercial plant for an extended period, which in petroleum refining is typically continuous operation for a year or longer before maintenance shutdown.

Specifically, if the catalyst was repeatedly brought into contact with continuous fresh biomass feed, with intermediate catalyst regenerations, as is necessary in evaluating the catalyst rate of deactivation and lifetime to determine its economic viability in a commercial refinery, Kelkar and coworkers would have found that the initial activity and selectivity of low-SAR ZSM-5 catalysts quickly decrease. This effect is more drastic as more biomass is introduced to the reactor and comes into contact with the catalyst particles. The rate of catalyst deactivation while on-stream for a refinery is one of the most important process cost variables. This was especially critical in KiOR's case because the inability to prolong the active lifetime on-stream of these catalysts, which are expensive at several thousand dollars per ton, negatively affected overall financial performance.

It becomes clear when reviewing the Kelkar example that more tests should be performed to correctly understand the changes in bio-oil yield and coke formation as more biomass feed is added and mixed with the catalyst, especially as biomass is converted over a long period of time on-stream. The extra testing would give a reliable sense of the catalyst deactivation rate for each catalyst sample being evaluated. Otherwise, the technology could take a wrong turn and head down a road to poor results, as will be discussed later on in context of KiOR's commercial biorefinery efforts.

Fortuitously, other researchers have made such measurements using HZSM-5 containing catalysts, which are ionic ZSM-5s paired with a hydrogen ion. These materials are prepared either by ion exchange of the freshly synthesized sodium zeolite with mild acidic water solutions, or more preferably by first ion-exchanging the material with an ammonium salt followed by a low-temperature calcination step to convert the ammonium ions to hydrogen ions and residual ammonia gas, which is removed. The sodium ions, originating from sodium aluminate used in the zeolite synthesis, must be removed because the sodium poisons the catalytic sites and reduces the catalytic activity.

In one informative example of HZSM-5 use, Shaolong Wan and colleagues describe the effects of different zeolite SARs on catalyst activity and selectivity [32]. They conclude that the larger number of acidic sites present in low SAR HZSM-5s leads to more rapid catalyst deactivation rates, as demonstrated by decreasing product yields. An HZSM-5 catalyst with SAR 25 exhibited a higher initial short-lived activity but quickly deactivated on-stream, whereas a catalyst with SAR 40, although it had a lower initial activity, exhibited a substantially slower rate of deactivation with higher activity and selectivity retention over a longer period on-stream, making the biomass conversion process much more cost effective.

Microporous (diameters less than 2 nm) and mesoporous (diameters between 2 and 50 nm) versions of these pentasil silicalite-type zeolites containing transition-metal ions or sometimes lanthanide-metal ions exhibit similar generic thermal, hydrothermal, catalytic, and deactivation performance as the low SAR HZSM-5 zeolites discussed above. For example, an iron ion-exchanged FeZSM-5, although it has a higher fresh deoxygenation activity and selectivity in upgrading crude bio-oil, deactivates on-stream at a higher rate than its precursor HZSM-5 when it is recycled between the reactor and regenerator, curtailing its longevity.

WHY SAR MATTERS

Silica-to-alumina molar ratios (SARs) play an important role in refinery catalyst selection. Less costly low SAR (2.2 to 3.6) wide-pore faujasites are relatively hydrothermally unstable and have high rates of activity loss on-stream in FCC units. Higher SAR (above 4.5) faujasites containing a smaller number of acidic aluminum active sites have better hydrothermal stability. Historically, these materials were arbitrarily separated at SAR 3.6 for no particular reason and incorrectly considered two different materials, coded Zeolite X and Zeolite Y, solely for patenting and marketing purposes.

The catalytic deactivation process relates mechanistically to zeolite crystallographic lattice transformations. During catalyst recycling, the zeolite crystal lattice reorganizes to a more energetically stable structure, and as it does so expels the iron cations, forming iron oxide moieties such as ferromagnetic Fe_2O_3 and lattice aluminum defects. These structural changes have been identified by electron spin resonance spectroscopy [33]. Furthermore, similar to FeZSM-5, thermally induced lattice transformations have been observed in ZSM-5 zeolites exchanged with other catalytic transition metals, such as nickel and copper.

Of importance to KiOR, these catalyst performance behaviors were validated in the pilot and demonstration plants and more importantly at Columbus I, the company's 500-ton-per-day commercial facility, using a commercial grade low-SAR ZSM-5 for converting low-metal-content woody biomass to liquid transportation fuels; more on the importance of biomass indigenous metal content later. Catalyst deactivation occurred quickly, requiring frequent large and costly replacements with fresh catalyst, rendering the overall commercial operation technoeconomically unfeasible.

For historical context, it is of interest to mention here that the challenging situation regarding the role of the crystallographic SAR composition of zeolites on their physicochemical and catalytic properties first cropped up in the early 1960s, when the first synthetic wide-pore faujasite-type zeolites were discovered and commercialized. Some signals that SAR affected thermal and hydrothermal stabilities came out in studies at Union Carbide's Linde Division on using zeolites in solid-state electronic devices, such as moisture sensors and high-temperature batteries, with improvement shown on using higher SAR material [34]. These researchers, which included Stamires, identified the stability issues as an inherent generic zeolite compositional problem; when working on applications of synthetic wide-pore faujasite zeolites as petroleum-refining catalysts, they observed the thermal stability problem of the low SAR zeolites, such as the so-called Sodium X type and its metal ion-exchanged forms. Substantial improvement was made by moving to higher SAR zeolites, above SAR 4.0 [35], and using these more stable zeolites to prepare cracking and hydro-cracking petroleum catalysts.

Of further note, the large-pore synthetic faujasite zeolites have different chemical composition, lattice structure, and pore archi-tecture than the small-pore pentasil silicalite-type ZSM-5s. Their ion-exchanged and ultrastabilized, or USY, forms, originally called "decationized" or "stabilized" zeolites, are described in detail along with the mechanism of their formation in the literature [36]. They are incorporated into cracking and hydrocracking commercial catalysts that are currently being used globally for petroleum refining, which also will be discussed in more detail later.

To put a punctuation mark on this discussion, the less costly low-SAR (2.2 to 3.6) wide-pore faujasites are relatively hydrothermally unstable, and as originally used had high rates of activity loss on-stream in refinery FCC units. The problem was resolved by moving to a higher SAR (above 4.5) faujasite containing a smaller number of acidic aluminum active sites and thus having better hydrothermal stability. Historically, this family of synthetic materials consists of a continuum of isomorphous polycrystalline compositions, which were arbitrarily divided at SAR 3.6. This demarcation was without any scientific basis, but it enabled patenting and marketing them in a somewhat misleading way as two different materials, Zeolite X and Zeolite Y, by Robert M. Milton in 1959 and Donald W. Breck in 1964 at Union Carbide [37,35].

The point here is that the role of the zeolite crystallographic SAR, both in the wide-pore faujasites and in the small-pore silicalites, is directionally the same—higher SARs lead to a slower rate of ther-mal deactivation and higher activity over a longer time on-stream. It should be noted that catalyst manufacturers prefer to produce and sell low-SAR wide-pore faujasite zeolites and low-SAR small-pore pentasil zeolites because they involve simpler production processes and lower cost raw materials than their high-SAR counterparts. We also note that the SAR designations have caused some confusion among zeolite makers, their customers, and researchers over time, given that large commercial production leads to materials over a wide SAR range.

2.2 DUE DILIGENCE: IMPORTANCE OF ESTABLISHING A PERFORMANCE BASELINE

Returning to KiOR's rush to commercial scale, in September 2008 the company's management decided to test the BCC technology further at a larger throughput capacity CFB pilot plant in Houston operated by petroleum engineering services company KBR. The KBR test was intended to duplicate the CPERI tests using pretreated biomass. As a catalyst, KiOR would use HTC, which had been used in the CPERI and ITQ tests.

In preparation for the test, the KiOR technical team proposed that the KBR reactor first be checked out and calibrated to establish a performance baseline and to confirm that the pilot plant was working properly, because the KBR reactor had never been used for thermocatalytic biomass conversion. The team expected that the calibration and optimization of process variables would allow the test to show conclusively whether KiOR's BCC technology, including biomass pretreatment and using the HTC catalyst, provided any advantages or disadvantages over the classic, well-established simpler approach involving thermolysis of biomass by hot sand in a CFB reactor. However, in a surprise to KBR and to the KiOR technical staff, O'Connor declined the calibration requests as he had earlier, and the KBR pilot-plant test was conducted blindly. In November, Loezos, one of the technical team members and an experienced oil refinery chemical engineer, reported that the test produced disappointing and unacceptable results.

Loezos, together with De Deken before he departed, had been analyzing and comparing published data by KiOR's competitors on using thermolytic or thermocatalytic pathways to convert biomass to liquid transportation fuels. Their technology comparisons and associated cost estimates convinced them that the BCC technology tested at CPERI, ITQ, and KBR was consistently not working or going to work. Moreover, at that time KiOR was running out of money.

Adding fuel to the flames, KiOR's management decided against advice from the senior technical staff to conduct an expensive commercial-scale trial of the BCC technology at Ivanhoe Energy's oil refinery in Bakersfield, California. The Ivanhoe facility had a throughput capacity of 1,000 barrels per day of heavy crude oil, converting it to light hydrocarbons. The plant at that time was being modified to process 15,000 barrels of crude oil or bio-oil per day using a catalyst regeneration unit with an inventory of 8 tons of sand or an FCC-type catalyst. But like the KBR facility, the refinery had never used biomass as feedstock. In addition, KiOR was planning, against advice that was given, to replace the sand with a proven track record of cracking heavy oil with the BCC HTC catalyst. However, after negotiations involving the need to perform costly equipment modifications and additions to handle biomass and feed it to the reactor, the Ivanhoe trial was canceled.

Besides declining to replace the untenable BCC technology, O'Connor as CTO further declined to allow the R&D staff to conduct calibration test runs to produce bio-oil at KiOR's own new pilot plants (Figure 6) and to use sand as the heat-transfer medium before starting to test new higher performance and lower cost catalysts. These tests would have at least allowed the team to find out if the pilot plants were working properly and were generating reliable and reproducible results. The tests would also have provided baseline data to compare with published data from CPERI and from competitors to help ascertain whether the BCC technology had any potential of being economically feasible, and beyond that, to be competitive with other leading related technologies.

Figure 6. KiOR's BCC pilot plant could process 2 kg of dry biomass per hour into crude bio-oil; the riser reactor is the vertical component at left.

As will be discussed further on, some postmortem published accounts of KiOR's demise have suggested that the reasons for declining the calibration and new catalyst evaluation tests may have been related to the cost and extra time it would have taken, not to mention that it likely would have exposed the failing BCC technology that the company's leadership was shielding from investors and the public. It must be re-emphasized that this situation later cost the company a substantial amount of money and wasted crucial time anyway, when without baseline data the R&D team struggled to correct the fundamental technology problems. While the above restrictions were playing out, KiOR under O'Connor's direction was continuing to fund the European contract labs for research work on the underperforming BCC technology. To that end, Ditsch, as Vice President of Strategy, was carefully following the R&D work for use in his fundraising efforts. After Ditsch reviewed the progress reports received from the outside labs, he concluded that their work was of no value and KiOR was wasting its money, and in October 2008 proposed to terminate the outsourced lab funding.

As if the company's situation was not precarious enough, at the end of 2009 KiOR's management faced another daunting problem—substantial legal costs for protecting and defending its intellectual property, when the prospects of getting new investor funding were growing slim. KiOR needed to protect its intellectual property, although underperforming and still under development, because it was the only asset the company had to keep operating. Potential investors needed to be convinced that KiOR's position was solid, having demonstrated on pilot-plant scale a technoeconomically favorable proof-of-concept that was properly patented to cover biomass demineralization pretreatments, design of CFB/ FCC-like reactors for bio-oil production, and new developments in hydrothermally stable zeolite catalysts. In effect, the company was counting on obtaining more money to buy time to work out the technical difficulties.

One example in late 2009 involved a former BIOeCON and KiOR consultant, George W. Huber, who was a postdoctoral researcher with Corma at ITQ. Huber had participated in company R&D discussions and had become knowledgeable of both companies' technology, patents filed, and new patent applications being prepared. Huber terminated his consulting contracts in August 2008 and moved on in his career as a chemical engineering professor at the University of Massachusetts, Amherst.

However, Huber had filed for his own patent in March 2008, while he still under contract and being compensated by KiOR. The patent, "Catalytic Pyrolysis of Solid Biomass and Related Biofuels, Aromatics, and Olefin Compounds," was issued in September 2009 and was assigned to UMass [38]. Huber's patent includes descriptions of processes, materials, catalysts including zeolite modifications, biomass demineralization pretreatments, and application claims similar to those of KiOR's intellectual property and patent applications. In November 2008, Huber formed his own company, Annellotech, which subsequently received a license from the university to commercialize the technology.

KiOR challenged Huber/UMass in court, and they came to an agreement to settle the lawsuit [39], defining a clear separation of potential business areas to avoid future competition and patent infringement. KiOR received the rights to use the technology for producing bio-oil and deoxygenated bio-oil to make gasoline, diesel fuel, jet fuel, and heating oil. Huber/UMass, and thus Anellotech, retained the rights for producing specialty chemicals, including aromatic compounds such as benzene, toluene, and xylene (BTX). Anellotech has since been successful in demonstrating viability of its BTX technology using wood or waste plastic at lab and pilot-scale plants. Anellotech and several other companies are now preparing to ramp up to commercial BTX facilities using biomass, sugars, plastic waste, and other feedstock materials [40].

3.0 COMMERCIAL PROCESS DEVELOPMENT CHALLENGES AND CORPORATE INTEGRITY

In late 2009 to early 2010, KiOR was in dire straits for not having a viable technology and no funds to continue operating. This situation came after the company ran through more than $100 million in its first two years focusing exclusively on the failing BCC technology. This had led to a division of key company personnel and formation of individual small teams that began looking for answers by working discreetly on different aspects of the project. In addition, some personnel left the company, and others were dismissed—the atmosphere in general was depressing and morale was low.

One team of senior managers, including Cannon, the President and CEO; Ditsch, the Vice President of Strategy; John Hacskaylo, Vice President of Research & Development; Christopher A. Artzer, Vice President and General Counsel; and Mitchell E. Loescher, Vice President of Technology, worked secretly from the other members of the management team in an effort to keep the company afloat. Separately, the stealth team made up of scientists and engineers, including Brady, Bartek, Zhang, Dydak, and Stamires, decided to work more diligently outside regular office hours, on holidays, and weekends to save the company by unraveling the technology problems and devising a viable new technology, with their main concern that the 10-ton-per-day demonstration plant and the Columbus

500-ton-per-day commercial plant would end up with the wrong process and wrong type of reactors.

The senior management team's approach included promoting KiOR's financial performance outlook by overstating existing data of bio-oil yields and understating fuel production costs. The group hoped to attract new investors and buy time to get the wrinkles out of the technology. Originally, members of this team stated bio-oil yields of 67 gallons per ton of dry biomass and a production cost of less than $1.80 per gallon of gasoline or diesel fuel. The executives later used these values in April 2011 in an application to the U.S. Securities & Exchange Commission (SEC) for an Initial Public Offering (IPO) of stock, leading KiOR to join NASDAQ in June 2012 for public trading. Subsequently, these publicly stated values were inflated further.

The management's agenda worked well to help create market capitalization of close to $2 billion by midsummer 2011 on expectations of close to $20 per share. Further boosting KiOR's business position, on June 22, 2011, former U.S. Secretary of State Condoleezza Rice joined KiOR's Board of Directors; Rice's appointment followed former British Prime Minister Tony Blair joining Khosla Venture's Cleantech Board of Directors in 2010 as a public policy advisor for clean technology companies. In addition, on the financial side, KiOR received term sheet papers from the U.S. Department of Energy in February 2011 with a $1 billion loan guarantee for the biofuel project.

At this time, a set of what in hindsight can be called a series of unfortunate events took place. William K. Coates had been hired on June 6, 2011, as a Vice President and Chief Operating Officer, coming to KiOR with more than 25 years of executive experience at major oil companies. Soon after he arrived at KiOR, Coates reviewed the information disclosed in the SEC application for the IPO. The documentation included an evaluation report of KiOR's technology performance prepared in 2010 by CPERI's Vasalos and Stephen McGovern, another consultant with experience in hydroprocessing and hydrocracking, as well as other company-reported technical information disclosed to investors. Subsequently, Coates spoke with laboratory technicians and pilot-plant operators to collect bio-oil yield data and related production cost details.

Finding that these measured data were substantially lower than the company's official statements let on, Coates discussed the data with Director of Finance Max Kricorian. Coates asked Kricorian to use these measured data in the company's financial model, which showed an untenable production cost of fuel. Surprised by the discrepancies in the figures, Coates called a special executive meeting with the management team on July 15 and informed those present that he had reviewed the available data, including the independent evaluation, and had concluded that the management had literally "cooked the books" and that he would inform the Board of Directors, and further that he was not going to be "part of this scam."

Coates soon after contacted Stamires and asked him to form a technical task force of in-house and outside experts to review options available to replace the failing BCC technology to avoid further entrenching its use. On Sunday, July 17, Stamires called Coates and informed him that he had organized the task force. Stamires had prepared a preliminary action plan to discuss with Coates and was ready to start the project immediately. Accordingly, Coates and Stamires arranged to meet in Coates's office on Monday, July 18, at 10:00 AM. However, one hour earlier, at 9:00 AM, Coates was visited at his office by Cannon and Artzer, who informed him he was being let go, after only five weeks on the job.

Regarding the formation of the stealth team of scientists and engineers, they were driven by O'Connor's declining to calibrate the performance of the pilot plants using sand and subsequently declining to test promising new catalysts outside the BCC technology. An indication of the seriousness of the concerns over KiOR purchasing the wrong equipment for the demo plant and potentially for the Columbus commercial plant is captured in an earlier discussion between Bartek and Stamires in March 2009. Stamires had asked Bartek to arrange with Ron Cordle, the pilot-plant supervisor, to work with his team during the weekends to calibrate the pilot plant using sand as the heat-transfer material to establish a performance baseline and to test some new catalysts in preparation for changing the reactor and process design. Stamires went so far as to offer to compensate Cordle and the pilot-plant operators for their extra time out of his own pocket, because the 2009 KiOR R&D work plan prepared by O'Connor focused only on the original BCC technology and did not include these kinds of projects.

This secretive overtime work later enabled the stealth team to achieve its objective of developing a replacement technology that would substantially increase the bio-oil yield and improve its quality. This action further enabled the team to reliably compare KIOR's BCC technology test results with those published in the open literature by competitors and with those from round-robin tests by Vasalos and Lappas using their CPERI pilot plant, with the same biomass, catalyst, sand, and operating conditions. This comparative work enabled KiOR to develop the new catalysts and gain confidence that their results were reliable and meaningful. Along the way, in February 2012, Stamires stepped down from the KiOR management team and as a member of the Science Advisory Board, though the company retained him as a full-time consultant.

The stealth team's work, building on previously reported technology discussed above, involved devising a two-reactor system and preparing and testing new nonzeolitic, low-activity, low-cost catalysts made with high-heat-capacity materials for biomass to bio-oil conversion and upgrading [41-43]. The team compiled its results and recommendations into a report, "Proposal for Commercial Use of an Efficient, Cost-Effective, Integrated Process for the Conversion of Biomass to Liquid Fuels," which Stamires submitted to Cannon and KiOR's management team on October 25, 2012. Stamires further asked Cannon at this time to form a task force "Team Oil Yield" to replace the failing BCC technology. However, the management team expressed no interest.

As part of its efforts, the stealth team had provided CPERI's Vasalos with raw bio-oil yield data and process parameters from KiOR's pilot plants and demonstration units for him to analyze and provide an independent expert opinion. Stamires had further arranged for Vasalos to visit KiOR and meet Cannon on January 24, 2012, hoping that Vasalos would be able to convince Cannon to change the technology and try a new approach. However, Cannon did not act on Vasalos' recommendations.

The stealth team believed the new approach described in its proposal was ready to roll and capable of immediately replacing the BCC technology. The proposal included a description with mechanical drawings of a dual-CFB reactor assembly set side-by-side or vertically upright. The close connection of the two units, the first of which could be an ebullated fluidized bed type and the second an FCC type, was designed to minimize the nascent bio-oil vapor residence time in the transfer pipe. This setup would also minimize molecular and free-radical interactions and/or polymerization, which are processes that prompt formation of larger molecular species that are difficult to deoxygenate and also produce larger coke deposits (Figure 7).

Biomass thermolytic devolatilization would take place in the first reactor, with the crude bio-oil transferred and thermocatalytically cracked and upgraded in the second reactor using new nonzeolitic, dual-function catalyst/heat-transfer materials developed and patented by KiOR. Each reactor had its own catalyst/heat-carrier regenerator to operate independently, and the mixing zone and riser length of each reactor were designed to carefully control reaction temperatures and residence times. Conventionally, the operating temperature of the first reactor is lower than the second reactor for optimal biomass devolatilization and maximum bio-oil yield. The operating temperature of the second reactor is optimized to sustain efficient bio-oil cracking and deoxygenation to maximize yields of deoxygenated light hydrocarbons.

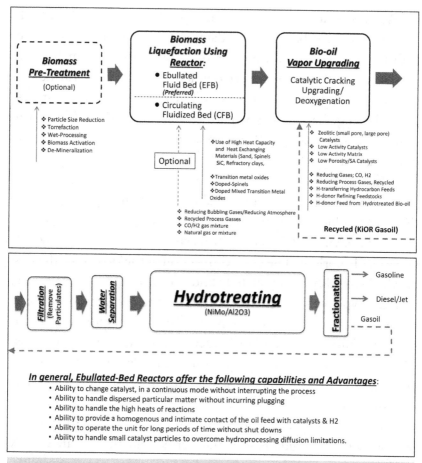

Figure 7. Design elements for a cost-effective integrated biomass conversion process to maximize bio-oil yield (top) and upgrade the crude bio-oil to commercial fuel products (bottom).

This process technology had the advantages of using inexpensive low-quality/high-metal-content waste biomass feeds and low-cost catalysts, with a bonus of requiring smaller volumes of hydrogen gas for bio-oil upgrading, to subsequently produce sufficient bio-oil yields and deliver refined gasoline, diesel fuel, and jet fuel. The approach included separating the char, ash, residual metals, and acidic aqueous phase from the bio-oil produced in the first reactor before upgrading it in the second reactor. The process was scalable to commercial-size plants handling 2,000-plus tons of dry biomass per day with a projected production of transportation fuels in the cost-competitive range of $2.00 to $3.00 per gallon, including delivery of the biomass to the production facility. The stealth development team recognized that this level of technoeconomic performance at large commercial facilities globally would go a long way in enabling renewable biofuels to make a difference. However, KiOR's senior management team did not act on the proposal.

In the meantime, the stealth team managed to introduce a new design of the reactor mixing zone in KiOR's existing CFB pilot plants and in the semiworks unit. The modifications increased the efficiency of the heat-transfer process from the catalyst regenerator unit to the reactor with a resulting increase in bio-oil yields. The team had also managed for the second time in collaboration with the CPERI team to calibrate KiOR's pilot plants using sand as the heat-transfer medium. The new test results were compared with published data of KiOR's competitors, and once again validated that the BCC technology was by far not efficient enough or cost effective and could not be competitive.

Furthermore, pilot-plant data obtained using biomass pretreated with added metal salts to improve thermolytic conversion in accord with KiOR's BCC technology demonstrated that this system produced low bio-oil yields. The supplemental metals, as well as metals naturally present in the biomass, interfered with catalyst activity. Coupled with the extra cost of biomass pretreatment and resulting metal-laden process water remediation and disposal, these findings unquestionably demonstrated that the BCC process was not going to work economically for KiOR.

In addition, the same pilot-plant data showed that using the HTC synthetic clay as a catalyst and heat-transfer medium produces excessive amounts of coke and gases and only a small volume of bio-oil. This information further validated that the BCC technology was not working or meeting KiOR's business objectives.

After O'Connor was removed by Khosla as KiOR CTO in 2009, the research team did make some progress in modifying the BCC technology, but it was not enough to completely solve the main problem. The changes involved eliminating the biomass pretreatment step and switching from the HTC catalyst to a commercial FCC-type ZSM-5 catalyst, overall a process resembling the approach CPERI used originally in 2002 [22] and also similar in some respects to the process used by Frankiewicz [13].

Regarding the pretreatment step, it is important to clarify what the term "pretreatment" really means, according to O'Connor, because in general this term has been used rather loosely in the published literature, including patents, for covering different chemical, thermochemical, mechanical, and other types of biomass conversion processes and combinations of processes. These include exotic pretreatments, for example, using high-energy electron beams to break down biomass chemical structure, as described by Xyleco, a company

TO PRETREAT OR NOT PRETREAT?

Raw biomass contains all the chemical elements needed to make fuels and chemicals, namely carbon and hydrogen, but a lot of elements not needed, such as oxygen, nitrogen, and metals. Plus, biomass is a recalcitrant biopolymeric material that is a challenge to breakdown and process at the molecular level. When approaching biomass conversion, decisions need to be made on process design to pretreat the biomass to make it easier to convert and/or remove unwanted elements, or to wait until later stages of the process to remove what is unwanted. The end goal is to create the overall most efficient, atom-economical, low-cost process that enables as much of the biomass to be utilized without generating waste or generating pollutants such as CO_2 and CH_4 emissions.

formed to convert biomass into value-added products. In this case, once the raw biomass is degraded by electron-beam bombardment, the material is further treated with enzymes to convert the cellulosic materials to sugars and further into alcohols such as ethanol. However, economic feasibility of this process at commercial scale has not yet been demonstrated. A story on Xyleco was broadcast by the television show *60 Minutes* on January 6, 2019 [44]. In a related technology, Bio-Sep is using ultrasound to produce cavitation (focused, high-pressure waves) that converts biomass into sugars, cellulose, and lignin for use in a range of products from specialty chemicals to building materials.

In KiOR's case, O'Connor's "pretreatment" term represents the technology described in his patent assigned to BIOeCON [45]. This process involves embedding small particles of inorganic materials/catalysts/metals into biomass particles before the composite formed is thermolytically or thermocatalytically converted to bio-oil. It should be noted that KiOR tested these "organic-inorganic" biomass-metal composite materials for producing bio-oil during 2008 and 2009 when O'Connor was KiOR's CTO. The results showed consistently low bio-oil yields and high volumes of coke and noncondensable gases being produced—this type of pretreatment is counterproductive.

Based on the present general understanding, these results are consistent with established physicochemical mechanisms indicating that the biomass indigenous metals or metals introduced by insertion into and/or by coating the biomass particles, as described in the O'Connor patent, suppress the primary devolatilization reaction that produces bio-oil as they promote secondary reactions such as gasification, at the cost of bio-oil yield. Based on this information, extensive R&D work during the past few years has aimed to develop cost-effective processes for removing/demetalizing the biomass indigenous metals before devolatilization. In the absence of the indigenous or introduced metals, the bio-oil yield increases while noncondensable gas production decreases.

In the end, KiOR's pilot-plant test results using the modified technology showed an increase in the bio-oil yield with acceptable oxygen content and lower coke and gas production, which was a substantial improvement compared with the original BCC process. But it was still by far short of being a technoeconomically feasible and scalable-to-commercial plant approach, as was subsequently validated by the failure of Columbus I.

3.1 CATALYST DESIGN AND DEVELOPMENT

It is important to note, as part of the lessons learned from KiOR, based on research by Stamires at the University of Cambridge with William Jones and colleagues [30], that calcining HTC at high temperature alters its crystallographic structure. The result is formation of a dense magnesium/aluminum spinel-type refractory material with a mesoporous surface area less than 50 m^2/g, and preferably less than 20 m^2/g.

When tested in a CFB pilot plant for converting biomass to bio-oil, this low-activity, high-heat-capacity, and high-heat-conductive refractory mixed-metal oxide basic catalyst produced less coke and gases and thereby substantially more bio-oil containing a reasonable amount of oxygen compared with its precursor crystalline HTC [41,43,46-47]. Furthermore, and more important, the overall thermocatalytic performance of these basic mixed-metal oxide compositions, including when improved by doping the precursor clays with catalytic metals such as zinc, iron, gallium, and others, is substantially more efficient for biomass devolatilization than the results obtained using the acidic ZSM zeolite-based commercial grade catalysts under the same conditions. This improvement was verified in KiOR's pilot plants. Upon calcination of a magnesium-rich spinel mixed-metal oxide composition, resulting in an increase in the Mg/Al molar ratio, a further increase in low-oxygen bio-oil yield was observed.

Although this finding for making spinel-like catalyst materials and metal-doped materials by thermally destroying the costly crystalline HTC precursor is interesting from an academic point of view, it is of limited practical commercial value, at least for biofuel production. The root cause is that the approach involves expensive pure magnesium and aluminum chemicals and processing to produce the precursor crystalline HTC, and then requires extra energy for high heat to calcine the material to destroy its crystalline structure—these technical challenges add to expenses.

Fortuitously, cost-effective thermal processes and low-cost raw materials are available for producing similar refractory spinel-like mixed-metal oxide compositions, for example, by using low-cost commercially available minerals. Among these options, bauxite or gibbsite provide an aluminum source, magnesite or periclase a magnesium source, dolomite is especially useful as a magnesium/calcium source, and lime is a calcium source.

Commercial grade magnesite ($MgCO_3$) has been used as catalyst in the thermocatalytic conversion of biomass to liquid hydrocarbons [48]. Additionally, lower cost magnesite ore, which contains calcium, iron, and silica impurities, besides being used in a variety of catalytic applications, has been used cost-effectively for the synthesis and production of layered double hydroxides, preparation of hydrotalcite-like anionic clays used as antiacids, synthetic catalytic reactions, and as biocompatible carriers/nanocarriers for medical drug-delivery systems. With their variety of commercial uses, synthetic and production technology processes have been described in several patents [30,49-52].

Periclase (MgO) is of particular interest because it also contains some iron along with smaller amounts of nickel, cobalt, and manganese to bolster catalytic activity. In fact, all these minerals contain some amounts of highly active catalyst metals and could actually be used directly without the need for diluting them with inert material and a binder to make catalyst particles. This would be a low-cost approach to thermocatalytic biomass conversion with the promise of increasing bio-oil yield compared with commercial zeolite-containing catalysts.

However, when used undiluted these minerals are catalytically quite active and produce excessive amounts of initial coke and gases and are chemically unstable during long process runs. Dispersing the minerals in microsphere particles mitigates these problems. As an example, for producing low-cost commercial catalysts with dual catalytic and/or heat-transfer properties for biomass conversions and bio-oil upgrading, Xiaodong Zhang and coworkers have reported an Fe(III)/CaO catalyst [53]. This catalyst, made by impregnating calcium oxide powder with iron nitrate solution followed by calcining, exhibits cracking activity mediated by iron and decarbonylation and deoxygenation activity mediated by calcium. It performs satisfactorily to produce reasonable yields of bio-oil with low oxygen content.

Still other low-cost options are available for developing dual-functional catalytic/heat-transfer materials for efficient cost-effective production of bio-oil, biofuels, and biobased chemicals. For example, the refractory minerals can be mixed with red mud [40,54]. Red mud is an inexpensive (essentially free) iron oxide-based waste material—effectively a clay-type mineral containing an assortment of metal oxides—generated during the Bayer process for producing alumina from bauxite ore. Red mud, somewhat similar to periclase, exhibits useful catalytic properties and high thermal conductivity, and it can be fused with magnesite ore to form mixed-metal oxide catalysts or with dolomite to create a lower cost version of Zhang's catalyst.

To that effect, an informative article in *Chemical & Engineering News* by Stephen Ritter (a coauthor of this book), "A More Natural Approach to Catalysts," explored red mud among other natural materials and by-products such as coal fly ash or biomass gasification ash as catalysts [55]. As reported, these globally abundant low-cost mineral wastes contain sufficient catalytically active metals to enhance biomass conversion. For example, Foster A. Agblevor and coworkers concluded that red mud potentially could replace commercial refinery catalysts such as ZSM-5s or other expensive hydrotreating/deoxygenation catalysts in a commercially feasible economically sustainable process [56-57].

At KiOR, the stealth team also explored red mud as a potential catalyst. But the team determined that red mud used by itself as a catalyst/heat carrier in biomass conversion to bio-oil would not perform satisfactorily, primarily owing to its high catalytic activity producing excessive amounts of coke and gases and little bio-oil. However, the team concluded that it could be incorporated in small portions, for example in low-cost clay-based microspheres, for catalytic applications such as deoxygenation of crude bio-oil with slower rates of deactivation and longer lifetime on-stream. The material also could be used to prepare spherical particles of refractory metal oxides when derived from pure materials, the aforementioned raw minerals, the Zhang iron-calcium catalyst, or made at lower cost from dolomite or lime doped with red mud, with optionally designed pore architectures suitable for doping with other catalytic metals.

For example, by varying the ratio of magnesite to bauxite in a fused mixture, a variety of refractory compositions are possible with the general formula AB_2O_4, where A can be a divalent metal ion such as calcium, magnesium, iron, nickel, manganese, or zinc and B can be a trivalent metal ion such as aluminum, iron, chromium, or manganese, as in $MgAl_2O_4$. Furthermore, these compositions can be doped by selected rare-earth metals (the lanthanides plus yttrium and scandium) to increase the heat conduction [58] and catalytic activity. Besides being used as catalysts on their own, they can also be used as additives when preparing catalyst particles to aid in removing sulfur and nitrogen from crude bio-oils for meeting environmental standards. Some of these have in fact been used commercially in refineries as FCC additives for reducing sulfur and nitrogen in gasoline, diesel fuel, and jet fuels.

It must be mentioned though that the above mixed-metal oxide spinel-like compositions can be used in thermocatalytic bio-oil production and upgrading without containing rare-earth metals. This is in contrast to typical FCC-type catalysts that contain trace, yet significant, amounts of rare-earth metals and are used by refineries in petroleum cracking. The absence of rare earths in bio-oil upgrading catalysts provides an advantage because a majority (roughly 60%) of rare earths used in a host of electronic devices from cell phones to computers and used in automobile catalytic converters are currently produced in China, which uses its market dominance as a geopolitical tool to control availability and prices, a situation that is fluid and could deteriorate. On the petroleum refining front, this type of "energy cold war" adversely affects global production and availability of transportation fuels, and also impacts global greenhouse gas emissions.

These mixed-metal oxide spinel-like compositions can be optimized by doping or impregnating them with other catalytic metals, including earth-abundant metals cobalt, nickel, molybdenum, and tungsten, and calcined to fix the metals on a γ-alumina support. The metals provide catalytic hydrogenation/deoxygenation and hydroconversion activities. These new high-performance hydrotreating catalysts are especially useful when the biomass feed contains waste plastics, which during thermolytic conversion generates hydrogen that interacts with the metals to initiate the hydrogenation/deoxygenation and hydroconversion reactions to reduce the bio-oil oxygen content. A further improvement in activity, selectivity, and regeneration efficiency is obtained when the γ-alumina support, obtained by dealuminating low-cost calcined kaolin clay and preparing an intermediate sol-gel, exhibits a calibrated architectural macroporosity supplementary to its mesoporosity [59-60]. Silicon carbide is another interesting material that has shown promise for bio-oil production for its high heat capacity and conductivity, and it is manufactured by fusing sand particles with rice husks or petroleum coke.

Another example of extensively researched materials—at a high cost, though—are the exotic synthetic mesoporous MCM-41 materials and their derivatives. These materials are interesting from an academic point of view, but once again, they are not for commercial use in economical thermocatalytic conversion of biomass to commodity transportation fuels. The MCM-41 materials are synthesized by dissolving the expensive crystalline ZSM-5 in concentrated sodium hydroxide solution, then with addition of a surfactant (CTAB) the solution is hydrothermally aged by autoclaving it for 20 hours. Following a pH adjustment, the material is autoclaved another 20 hours and finally filtered, washed, and treated with an ammonium salt for ion exchange—overall an expensive process [61].

The bottom line here is that these mesoporous materials are more costly than their precursor ZSM-5 and they are less hydrothermally stable and still produce small volumes of bio-oil. However, the cost problem can be substantially reduced by using cheap minerals such as diatomite or kaolinite and/or dealuminated kaolinite and partially converting them to zeolite materials, that is clay-based microspheres containing mesoporous ZSM-5. KiOR's research demonstrated using some of these inexpensive minerals to produce refinery catalysts with customized pore architectures suitable for upgrading crude bio-oil to light deoxygenated hydrocarbons for transportation fuels and commodity chemicals, such as BTX and propylene.

One example of a plausible process to meet the need to make these low-cost FCC-type ZSM-5 microsphere catalyst particles is to start with kaolin clay, which is available globally in large quantities. For this type of catalyst preparation, low-grade off-white material containing small amounts of metal impurities such as iron, sodium, magnesium, calcium, and titanium is acceptable. The content of these metals in kaolin varies among sources globally.

In the first processing step (Figure 8), the kaolin is calcined at 850-900 °C to dehydrate the aluminosilicate material and form a denser metakaolin clay phase. In the second step, a portion of the metakaolin is treated with sodium hydroxide solution to extract silica, producing a sodium silicate solution and desilicated clay. Subsequently, these two products are used in formation of the catalyst particles. A part of the sodium silicate solution can be used as silica gel for the crystallization of the zeolite crystals, and another part of the solution can be passed through a hydrogen ion exchange column to produce polysilicic acid to use as a particle binder to form the catalyst particles.

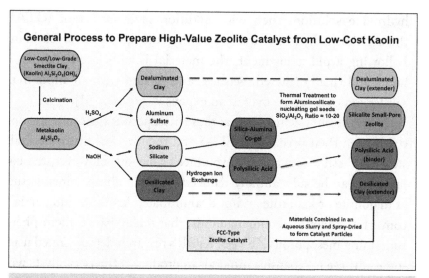

Figure 8. General Process to Prepare High-Value Zeolite Catalysts from Low-Cost Kaolin. A technoeconomically feasible biomass conversion process will require an inexpensive FCC-type catalyst. One plausible approach is to start with low-cost kaolin clay, with low-grade off-white color material containing small amounts of metal impurities such as sodium, magnesium, calcium, and titanium being acceptable. Additional cost savings could be gained by including by-product clay minerals/ mining wastes. Following calcination and then acidic or basic processing to prepare needed intermediates, sodium silicate and aluminum sulfate are combined to form a silica-alumina co-gel in proportions to achieve the desirable silica-to-alumina ratio (SAR). This co-gel can be transformed to "amorphous aluminosilicate nucleating gel seeds" with a SAR in the range of 10 to 20. Combining these seeds with additional sodium silicate and ammonium sulfate, the formed alkaline slurry is aged to synthesize aluminosilicate zeolites having a sodalite crystallographic structure, such as the ZSM-5s. In a final step, a homogeneous aqueous slurry is prepared with the crystalline zeolite, dealuminated/desilicated metakaolin serving as an extender, and polysilicic acid serving as a binder. With pH adjustment, the slurry is spray-dried to form catalyst particle microspheres. The zeolite before or when incorporated into the microspheres can be modified by treatment with acid, base, ammonium salts, phosphate salts, selected metals, and more, with intermediate thermal/calcination treatments.

A parallel processing step involves dealumination of metakaolin via an acid leaching process, using sulfuric, nitric, or hydrochloric acid. Using sulfuric acid, for example, leads to aluminum sulfate. Furthermore, the dealuminated metakaolin product can be used as a catalyst particle extender/matrix when compounding the catalyst particles. In a subsequent step, the sodium silicate and aluminum sulfate are combined to form a silica-alumina co-gel in proportions to achieve the desirable SAR. This co-gel can be transformed to "amorphous aluminosilicate nucleating gel seeds" with a SAR in the range of 10 to 20. Combining these seeds with additional sodium silicate and ammonium sulfate, the alkaline slurry is aged to synthesize aluminosilicate zeolites having a sodalite crystallographic structure, such as the ZSM-5s. The zeolite synthesis has been described in a 2003 patent assigned to AkzoNobel [62].

In a final step to assemble the catalyst particles, a homogeneous aqueous slurry is prepared with the dealuminated/desilicated metakaolin, crystalline zeolite, and polysilicic acid binder, with an optional pH adjustment, and spray-dried to form microspheres. The zeolite before or when incorporated into the microspheres can be modified by treatment with acid, base, ammonium salts, phosphate salts, selected metals, and more, with optional intermediate thermal/calcination treatments. In addition, mesoporous MCM-41 zeolites as described earlier may be synthesized using these low-cost components. Still further, low-cost hydrocracking and hydrotreating catalysts to upgrade the bio-oil to low-oxygen, low-sulfur, and low-nitrogen lighter hydrocarbons can be produced from the silica-alumina co-gel, or the alumina can be used alone as a catalyst metal support or catalyst extender after being calcined to form γ-alumina.

The essence of the Agblevor work regarding the use of red mud, the use of so-called natural catalysts, and the catalyst preparation example given above is that these teachings in general provide realistic examples pointing out the future direction, needed priorities, and basis for financial support by industry, investors, and government funding agencies for global R&D work to commercialize production of lower cost small-pore silicalite and wide-pore faujasite zeolite crystals along with catalyst particle binders and extenders. A key objective is lowering the catalyst selling price.

If indeed renewable biofuels will ever be able to overcome cost barriers to make a significant positive contribution in supplementing or even replacing petroleum-derived fuels and chemicals, we will need to use these low-cost material resources and minimal processing steps to develop commercial-scale biorefineries with processing capacity of 2,000-plus tons of biomass per day. This level of process intensity is needed to compete with the current production cost of petroleum-derived transportation fuels and chemicals, especially when petroleum is globally available and the price is low, below $60 per barrel. But the reality in 2022 was that West Texas Intermediate crude that is ideal for gasoline and Brent crude ideal for diesel—the standards for petroleum refining—reached as high as $123 and $128 per barrel in March before settling back down below $100. These price swings hint at an ongoing opportunity for biofuels. Otherwise, most R&D on renewable biofuels completed so far and what might be achieved in the future will remain only of academic interest, with no commercial utility, and result in financial losses of the contributions of taxpayers and investors.

3.2 COMMERCIAL REACTOR DESIGN

Now, returning to KiOR's activities, some progress had been made when the R&D staff, prompted by work of the stealth team, had managed to introduce, as a quick fix, small-pore ZSM-5 catalysts to replace the HTC synthetic clay. The ZSM-5 catalysts had been tested already at CPERI and showed higher bio-oil yields and less coke than the HTC catalyst, with reasonable deoxygenation of the bio-oil and reduced amounts of the undesirable by-products. Overall, this was a substantial improvement, but the bio-oil yield was still far short of what was needed and the catalyst cost too high.

The stealth team went further, experimentally validating the critical dual function of the catalyst particles. Besides providing catalytic sites for the chemical transformations to take place, the catalyst particles need to have high heat capacity and high heat conductivity. These properties allow the material to load up enough heat while in the catalyst regenerator unit, which operates at about 700 °C, and then quickly transfer the heat at a rate of about 1,000 °C per second to the biomass particles in the reactor unit, which operates at about 450-500 °C. Large, fast heat fluxes are essential for driving the thermolysis reactions going on in the reactor mixing zone, where the biomass particles have on average a residence time of about half a second, which in turn helps determine bio-oil quality and yield. The total amount of energy needed to heat the biomass particles to the "flash" devolatilization/thermolysis temperature and to drive the reaction is in the range of 1 to 2 MJ/kg of dry biomass.

The heat challenge comes from the biomass particles, even in small sizes, which generally exhibit low heat capacity and poor thermal conductivity, and from the porous, low-density catalyst particles, which also have low heat capacity and conductivity. The combined effects of these thermal inefficiencies result in incomplete biomass thermolysis, and therefore contribute to low bio-oil yield. In addition, using more energy for mechanical processing or to create heat for processing the biomass adds to production costs.

Combining the biomass thermolysis step with the bio-oil/vapor catalytic deoxygenation upgrading step into a one-reactor system represented an attractive concept in the early 2000s, at least for capital equipment cost-saving purposes, especially when using available oil refinery FCC reactor units. However, much of the R&D up to that point had been on improving catalyst activity and stability, and not much attention was given to improving the heat-capacity and heat-transfer properties of the catalyst.

The stealth team arranged for KiOR to license the 1 kg/hour throughput Vasalos-Lappas/CPERI FCC biomass pilot-plant reactor design (Figure 9A/B) [22,63-64]. This was the design used for

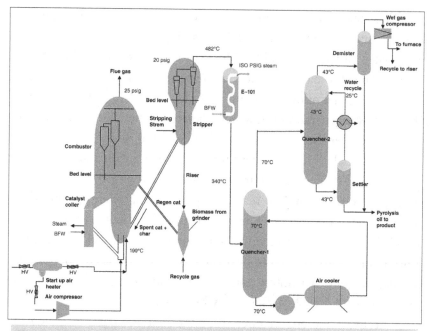

Figure 9A. Schematic diagram of an FCC-type pilot plant used by Vasalos and Lappas, and also by KiOR, as a process development unit for biomass conversion to bio-oil. Reprinted with permission from Ind Eng Chem Res 2008;47(3):742-747.

constructing KiOR's pilot plants, semiworks unit, and the Columbus I plant. As mentioned earlier, KiOR and CPERI performed round-robin collaborative tests with their respective pilot plants using the same high-quality, low-metals biomass and commercial grade ZSM-5 FCC catalysts, including using the same type of sand.

Of importance here was the design of the reactor, which included a mixing chamber at the bottom of the riser to provide more efficient intimate biomass/catalyst mixing and much faster heat transfer rates than were used traditionally. These reactor design improvements resulted in increasing the bio-oil yield close to 70% on a dry biomass basis, with less char, coke, and noncondensable gases. In addition, a shorter length riser minimized the residence time of the bio-oil vapor to less than two seconds to reduce overcracking and cross-molecular/

free-radical interactions and avoid larger molecule formation and coke production. Equally important were the subsequent processing steps, including achieving an efficient rapid separation of vapors from solids at the cyclones and efficient quenching of oil vapors, where the design optionally allowed trapping noncondensable gases such as CO, CO_2 and CH_4 as part of a decarbonizing strategy to prevent their release to the environment. These greenhouse gases in turn can be utilized in other large-volume chemical processes, for example to make building materials, or added to natural gas supplies similar to gas emissions from landfills. Process conditions typically were a reaction temperature of 450 °C, regenerator temperature 620 °C, and vapor residence time close to 0.5 seconds.

The bio-oil yields using sand and milled debarked pine wood chips were in the range of 70 to 80 gallons per dry ton of biomass with a high 30% to 40% oxygen content. The bio-oil yield using the ZSM-5 catalyst in the same pilot plants using the same biomass feeds was about a half of that produced by the sand, but it contained about half as much oxygen. These results were reproducible and confirmed in another pilot plant of comparable reactor size.

However, this performance is idealistic and short-lived, because the process deteriorates technoeconomically as the zeolite-containing catalyst deactivates quickly on long runs, producing less bio-oil with higher oxygen content. This catalyst problem is a result of low hydrothermal stability in an acidic environment, plus metal poisoning, along with the detrimental high heat of the regenerator. For example, in KiOR's demonstration unit with 10 tons per day dry biomass conversion using milled debarked pine wood and a commercial ZSM-5 catalyst with SAR 25, the continuous production of bio-oil containing 15% oxygen decreased by about 25% over about 10 days on-stream with continuous catalyst regeneration.

This limitation is greater with catalysts containing alumina-type particle binders, such as aluminum chlorohydrol or gelled pseudoboehmite alumina. Therefore, this particular problem is minimized by using a silica type particle binder, such as polysilicic acid or plain silica

Figure 9B. Vasalos, Lappas, and colleagues used this CPERI pilot plant for KiOR's early R&D work, which provided reproducible results on catalyst and overall process performance.

gel. Additionally, these zeolite stability and functionality limitations, and particularly so with low-SAR catalysts, are compounded with a heat-deficiency problem in the reactor, which becomes substantial as the size of the reactor mixer chamber is increased to increase throughput capacity for a commercial plant. KiOR demonstrated these shortcomings by comparing the performance of the different-sized reactors, concluding that scale up to commercial size, such as at the Columbus plant, becomes economically unfeasible.

It should be mentioned that improvement of the low-SAR zeolite's thermocatalytic performance is possible by increasing the crystal's SAR and also by phosphating the material before it is incorporated in the compounding water slurry with the binder and clay extender; alternatively, the phosphating can be carried out once the microspheres are prepared. Overall, these findings indicate that the usual commercial FCC-type microsphere catalyst particles containing a ZSM-5 are unsuitable for commercial plants using one-pot in situ biomass conversion reactors, which will be discussed later in more detail.

To minimize these performance limitations, the stealth team replaced the commercial ZSM-5 catalysts in the CFB pilot plant with its new low-cost nonzeolitic catalysts, which exhibit lower initial fresh catalytic activity but have lower deactivation rates when kept longer on-stream. These new catalysts produced less coke and char and higher bio-oil yields, with the bio-oil having an acceptable oxygen content, as compared with the performance of the commercial low-SAR ZSM-5 catalysts. Some of these new nonzeolitic catalysts are described in a 2014 KiOR patent [42].

3.3 HEAT TRANSFER DEFICIENCY, CATALYST LIMITATIONS, AND SOME REMEDIES

The crux of KiOR's technology problem, the stealth team had determined, was heat deficiency or "heat starvation" in the mixing zone of the riser of the biomass reactor. In other words, operating with an inefficient heat-transfer process. This deficiency increases in intensity with increasing throughput capacity and size of the reactor on moving from pilot plant through to commercial scale. It should be noted that volumetric-driven heat-transfer deficiencies cannot be identified when only using small benchtop rector equipment for biomass thermocatalytic devolatilization.

The stealth team therefore found that conventional porous, low-density FCC-type catalysts used in a dual role as catalyst and as heat-transfer medium in a one-pot reactor system cannot be scaled up technoeconomically to a large demonstration unit or commercial-sized plant, without substantially compromising bio-oil yield and profitability. Proof was provided by the measured bio-oil yields in the mid-50s gallons per ton of dry biomass at the company's bench-scale reactors, mid-40s at the pilot-plant scale, mid-30s at the semiworks/demonstration plant, and low-20s at the Columbus I commercial plant. These yields were measured and compared at 15% bio-oil oxygen content, the maximum practical level for upgrading in a typical refinery; 10% bio-oil oxygen content is more ideal, requiring less hydrogen and catalyst for subsequent hydrocracking and hydrotreating processing to make liquid transportation fuels.

In addition, pilot-plant tests to determine the catalyst deactivation rate on-stream over periods of several weeks further indicated that the bio-oil yield and oxygen content both increased with time. On the contrary, the bio-oil yield at the semiworks/demonstration plant decreased and the oxygen content increased with time. This pattern of decreasing bio-oil yield and increasing oxygen content was also observed at the Columbus commercial plant.

The heat deficiency in the reactor stemmed in part from insufficient heat maintenance and transfer via convection and conduction, and in part from the heat capacity and conduction properties of the chemical components of the catalyst microspheres. As noted earlier, enlarging the reactor volume when scaling up exacerbated this condition. A partial solution to this problem was sand. This ubiquitous material has a high heat capacity to maintain bio-oil yield on scale-up.

Although sand is used commercially as an efficient heat storage and thermal insulation material, it exhibits relatively low heat conductivity, with less-than-optimal heat transfer or exchange properties. This property limitation presents a disadvantage for thermolytic biomass conversion, wherein the heat must be quickly transferred to the biomass particles so they reach the volatilization temperature at the center of the particles while they are in the reactor—a process that needs to take place in less than half a second. The speed is vital, and the materials used are vital, to convert the biomass particles to a vapor via selective primary decomposition reactions while suppressing secondary reactions that produce undesired coke and gases. In general, any hydrothermally and chemically stable, high-bulk-density material could be used, provided it has low surface area and low pore volume.

Historically, since the 1940s, heat deficiencies at the reaction zone of CFB/FCC type conversion units with continuous catalyst regeneration have been a problem in petroleum catalytic cracking. Early proposed solutions included adding inexpensive, similar-sized microsphere particles made from high-heat-capacity materials, including sand, as noted in examples described previously in our story [65]. Researchers have found that including such materials in the reactor/mixing zone immediately increases the yield of light hydrocarbons. However, an undesirable side effect occurs: The hard sand particles, clashing with the softer catalyst particles and with the walls of the reactor and regenerator in the steamy acidic environment, causes the catalyst to abrade and crumble into fine powder—effectively a "sandblasting effect."

This fine powder could coat the reactor hardware and clog valves and be difficult to separate from the bio-oil, plus create air pollution when fine particulates escape and enter the atmosphere. For this reason, catalyst microsphere particles with high attrition resistance are preferred in oil refinery FCC units to avoid catalyst particle loss. Oil refinery process engineers who operate FCC units, such as those who went to work at KiOR, are experienced in dealing with these attrition problems. In some cases, oil refineries have no choice but to add softer catalysts or add blends of softer and harder catalyst particles to an FCC unit because of material availability. Ultimately, the use of catalyst plus sand blends, when the catalyst will be exposed to severe mechanical circulating treatments for long periods, is not an overall technoeconomically feasible operation.

One solution is incorporating heat-retention materials into the catalyst particles, such as magnesium borate, aluminum borate, boron phosphate, or mixed-metal oxides [66]. The two-particle composite system containing microsphere particles with similar attrition resistance achieves a dual function in the reactor mixing zone, serving as an efficient heat-transfer medium and as a low-activity catalyst to cause some cracking and deoxygenation of the crude bio-oil. The resulting bio-oil containing a smaller amount of oxygen, and being less acidic, would be more suitable for further cracking and deoxygenation in a second reactor, producing less coke and char and higher fuel yields while requiring a smaller volume of hydrogen. Accordingly, the overall production cost of biofuels and specialty chemicals such as BTX would be reduced substantially.

However, these materials, when exposed to low pH water/vapor and high temperature, also chemically break down and form reactive metal salts, leading to destruction of the catalyst active sites in the crystal lattice. Therefore, they cannot be used for in situ, single-reactor thermolytic or thermocatalytic biomass conversion to liquid hydrocarbons, such as KiOR was attempting.

That leaves sand as an option, as noted earlier, and being dead cheap, it is preferable, although it has less than optimal heat conductivity and no catalytic activity. To solve that latter problem, much R&D work has come to pass on small-pore pentasil zeolites such as ZSM-5s as catalysts for cracking and deoxygenation of crude bio-oil via decarbonylation/decarboxylation and dehydration/dehydroxylation routes. These zeolites, although they substantially deoxygenate bio-oil, lead to production of excessive amounts of by-products such as CO, CO_2, CH_4, and acidified water at the expense of bio-oil yield. Additionally, also as noted earlier, these catalysts are expensive, and they also deactivate faster than one would like in the presence of biomass indigenous metals and under the acidic hydrothermal conditions in the reactor, and even further during recycling at the higher temperature in the catalyst regenerator. Beyond frequent regeneration, this situation requires a high level of replacement with fresh catalyst, thereby rendering the overall commercial process economically unsustainable.

We also need to consider that while we can architecturally change the mesostructure of large- and small-pore zeolite crystallites to beneficial effect, for example to increase their catalytic product yields with optimized diffusion and transport of reactants and products, the formation of crystal mesopores decreases zeolite particle density as internal surface area increases and reduces heat capacity and heat conductivity. This counters the benefits of the mesostructure sculpturing. These two competing and possibly opposing effects can result in variable results on product yields and can lead to a difference in the quality and composition of the hydrocarbon products. To determine the outcome, one needs to experimentally obtain an optimized balance of the yield, composition, and quality versus size and amount of mesopores.

A remedy for the decrease in particle heat capacity and heat conductivity is incorporating microfine particles of an efficient heat-transfer agent with the zeolite powder when formulating the catalyst, as described earlier (Figure 8). Although heat deficiencies of the catalyst particles in small-volume fixed-bed and CFB pilot-plant reactors are difficult to detect and measure, when recognized the size and volume of mesopores should be optimized with respect to their desirable chemical composition and light hydrocarbon yield for cost determination purposes and then tested in long-run deactivation aging tests in large throughput CFB pilot units before building a commercial-sized plant designed to use mesostructured zeolite catalysts. More details on the intricate nature of zeolite pore structure are discussed later on.

Taking all these considerations into account, to this point in our story we can say that a better balancing act is still needed.

3.4 A TWO-REACTOR BIOMASS CONVERSION TO LIQUID FUELS SYSTEM

Continuing forward in thinking about a two-reactor system, one potential commercial cost-effective application using a ZSM-5 or other
catalyst might involve thermal biomass devolatilization in the first
reactor, which could be a fluidized ebullated bed reactor that uses
sand, crushed recycled glass, silica, alumina, natural clays, or another
efficient refractory high-heat-capacity and heat-transfer medium containing supplemental entities such as magnesia, carbon, calcium, or
rare-earth metals. In this approach, the metal-containing ash, char,
and acidic water/vapor would be removed in the first reactor. The
separated crude bio-oil/bio-oil vapor could then be transferred to a
second reactor, which could be an FCC type, for traditional catalytic
cracking and deoxygenation upgrading using a zeolitic or nonzeolitic
catalyst or a blend of a zeolitic or nonzeolitic catalyst with other
catalysts. These blends can provide a twofold action: the cracking
activity and selectivity of the acidic zeolite catalyst, and the deoxygenation activity of basic mixed-metal oxide and spinel-like catalyst
compositions.

For example, in the second reactor a 50:50 or 25:75 blend of
conventional ZSM-5 catalyst with a nonzeolitic material could be
used, with the catalyst microspheres of both types having the same
size, same particle density, and similar attrition resistance. KiOR
patents describe nonzeolitic microspheres that comprise the needed
dense, high-heat-capacity, low-catalytic-activity refractory materials
[40-42]. These blends can produce higher bio-oil yields with oxygen
content comparable with that obtained when using 100% ZSM-5
catalysts, resulting in substantial production cost savings.

To demonstrate the paramount importance of catalyst cost, the
case of the planned KiOR commercial trial of the BCC technology
at Ivanhoe Energy's oil refinery in Bakersfield, California, provides
an illuminating example. As mentioned earlier, the Ivanhoe facility
had a throughput capacity of 1,000 barrels per day of heavy crude

oil, converting it to light hydrocarbons. The plant had never used biomass as feedstock, though at that time was being modified to process 15,000 barrels of crude oil or bio-oil per day using a catalyst regeneration unit with an inventory of 8 tons of sand or an FCC-type catalyst. It turns out negotiations involving the need to perform equipment modifications and additions to handle biomass and feed it to the reactor broke down, and the Ivanhoe trial was canceled.

But in a hypothetical case, thermocatalytic biomass conversion to liquid hydrocarbons using the above-described blend of zeolite and nonzeolite catalysts at Ivanhoe would have required 2 tons of a zeolite-containing catalyst such as ZSM-5 costing close to $9,000 per ton plus 6 tons of a mixed-metal oxide catalyst costing close to $1,000 per ton. The total catalyst-charge cost would be about $24,000. Alternatively, using only a zeolite catalyst would cost $72,000. The potential cost savings would be $48,000, just for the initial catalyst charge, and not include the additional daily cost savings realized by

WHEN TWO IS BETTER THAN ONE

Biomass conversion to usable transportation fuels is inherently a two-step process: First converting the raw biomass into a processable oil, then refining the oil to remove impurities and isolate the desired hydrocarbon fuel components. While the two-step process could be carried out in a single reactor, the reaction conditions and array of by-products created in the initial step inhibit the efficiency of the second step, especially when it comes to longevity of the catalysts that are critical for producing the finished fuel in an economically viable way. The challenge then becomes selecting and formulating the catalysts and other materials needed to produce the bio-oil and upgrade it to the target fuel.

using the blend to replace the withdrawn spent catalyst to maintain activity and selectivity while on-stream.

Moving on, a variation in operating a two-reactor system could involve using the new low-cost, low-activity refractory materials in both reactors, with the option that the second reactor employ a material with higher catalytic activity. Or still alternatively, the material used in the second reactor could be a ZSM-5, a modified ZSM, an MCM-41, or other zeolite catalyst. Either way, deeper deoxygenation could be achieved by using the low-cost refractory materials at the front end of the process, which in turn would require less hydrogen and hydroprocessing catalysts during the second stage upgrading and cut production costs substantially.

Because it appears pentasil ZSM-5 zeolites would be a leading choice for commercial catalysts for upgrading crude bio-oil to light deoxygenated hydrocarbons for some time to come, and at present their high cost limits their commercial use, it is worth exploring further any effective processes that could reduce production cost. In this regard, the technology of growing zeolites on preformed clay-based microspheres is interesting. This technology was originally developed in the late 1960s by Engelhard Corp., now a part of BASF, a global FCC catalyst producer credited with developing the first automobile catalytic converter.

As described earlier, microsphere particles typically are formed by spray-drying a mixture of kaolin clay and silica binder, with the microspheres subsequently calcined at high temperature to convert the kaolin to a high-bulk-density spinel/mullite phase material. In the Engelhard/BASF technology, the pentasil ZSM-5 zeolite is then formed in the outer shell surface of these materials by treating the microspheres at controlled temperature and high pH in a water slurry containing low-cost amorphous alumina and silica seed material [67]. The microspheres are further ion exchanged and/or calcined and the microspheres further stabilized with additives. This approach has the benefit of incorporating the zeolite into the catalyst particles during the process rather than starting with the expensive ZSM-5 as the seeding precursor, which avoids costly organic templates and simplifies the ion-exchange process. Close to 500,000 tons of these catalysts along with faujasite zeolite Y catalysts are being produced annually worldwide.

Besides the cost advantage of these zeolite-clay composites, they also provide interesting and useful dual-function physicochemical properties along with catalytic activity and selectivity. Specifically, the dense spinel/mullite core of the microspheres has a higher heat capacity than that of conventional highly porous FCC catalyst microspheres. The calcined clay-based microspheres also have higher attrition resistance properties, and thereby create minimal amounts of problematic microfine particles during oil refinery service.

Furthermore, the process for preparing catalysts from preformed calcined clay-based microspheres provides the advantage of controlling the amount of the zeolite formed on the particles and hence controlling the particles' shell skin depth as desired. By limiting the thickness of the zeolite skin layer, the diffusion and residence time of the reactants and products inside the particle are limited, hence less coke forms, which in turn reduces the rate of catalyst deactivation, reduces catalyst replacement cost, and improves bio-oil production, making the overall process more cost effective.

3.5 BIO-OIL UPGRADING TO FUELS AND CHEMICALS

Bio-oil, like petroleum, is viscous and contains high-molecular-weight compounds containing varying amounts of oxygen, sulfur, and nitrogen that must be removed to prepare light hydrocarbons usable as fuel components and as chemicals. Petroleum is conventionally upgraded via hydroprocessing methods that employ specialty catalysts in fixed or ebullated bed reactors operating under a pressurized hydrogen atmosphere and at temperatures up to 400 °C. The catalysts are typically alumina doped with transition metals or with transition-metal oxides or sulfides.

In 1973, motivated by the oil crisis, researchers focused on developing high-performance hydrotreating catalysts and processes to convert shale oil and tar sands to transportation fuels. One critical point, for example with transition-metal γ-alumina, is that the catalyst particles must exhibit a macroporous structure (pores greater than 50 nm diameter). This pore size is needed to accommodate the large hydrocarbon molecules in crude oil from shale and tar sands, and now from bio-oil, that need to be hydrogenated when removing the oxygen, sulfur, and nitrogen, in particular for use in low and ultralow sulfur fuels. An example of this pioneering research was carried out by Alfred J. Toombs and Warren E. Armstrong, with a patent assigned to Shell Oil [68].

Toombs and Armstrong prepared a catalyst by adding a sucrose solution to gelled alumina, drying the material at 120 °C followed by heating the dried material at 400 °C to carbonize the sucrose within the alumina matrix. They further heated the "sweet" alumina to 1,300 °C in an oxygen-free helium atmosphere to form refractory aluminum oxide, and finally heated the material to 500 °C in air to burn off the carbon and leave behind the pores. The resulting macroporous alumina exhibits a surface area of 86 m²/g. By comparison, the same alumina without the sugar treatment exhibits a surface area of only 2.6 m²/g. Similar metal-support materials impregnated with catalyst metals have been subsequently invented using other solubilized carbon pore-forming materials, including starches, cellulose, carbon blacks, stearic acid, polyvinyl alcohol, polyethylene glycols, and polyacrylamides.

A technology to achieve desirable pore architecture with a calibrated macroporous structure was created in 1982 via a notable advance in the carbonization process. John Lim, Michael Brady, and Kirk Novak, in a patent assigned to Filtrol Corp. (a subsidiary of Kaiser Aluminum), describe a process in which a commercially available carbon black powder is selected with a uniform particle size and incorporated into a gelled-like acid peptized boehmite alumina matrix [59]. The designed macroporosity, typically in the range of pore diameters from 500 to 900 nm, supplements the mesoporosity in the range of pore diameters from 5 to 30 nm. This architecture increases the accessibility of the reactant molecules to the metal active sites, which increases product yield and further renders the hydrocracking and hydrotreating catalyst regeneration process more efficient and cost-effective.

The selected pore-regulating carbon black was used to produce hydroprocessing catalysts containing multimetallic phases made from a combination of metals, mostly earth-abundant nickel, cobalt, molybdenum, and tungsten. Some high-activity, high-metal hydroprocessing catalysts may contain up to half the weight of the dry catalyst as metal oxides. When these catalysts are used in an ebullated bed reactor, higher viscosity crude bio-oils containing larger amounts of oxygen can be hydrotreated and upgraded, as opposed to processing in fixed-bed reactors that frequently encounter the usual problems of pore plugging and process interruptions. This approach has the advantage of allowing the use of a less active catalyst and more selective replacement, with longer life on-stream for raw biomass devolatilization at the front end of the process. This technology was also used by these researchers for the preparation of specialty FCC type catalysts used for cracking heavy petroleum feeds.

Later on, when at KiOR, Brady and his colleagues replaced the costly commercial carbon black with waste biomass as a combustible carbon pore-regulating additive—finely ground micronized lignocellulosic material, the same low-cost feed used in commercial biomass conversion reactors to make bio-oil [69]. They started with pine wood sawdust milled to produce 1 μm or smaller particles and mixed with gelled alumina in a 5:95 weight ratio. This mixture was milled further to produce a homogeneous dispersion, extruded to form 1.6 mm pellets, dried at 200 °C, and calcined at 650 °C to burn off the dispersed wood carbon and form 50- to 500-nm pores in the alumina particles. With this macroporous support material in hand, the pores could be impregnated with selected catalytic transition metals and/or lanthanide metals as desired, and optionally phosphated. For example, conventional hydrotreating and hydrocracking catalysts for bio-oil hydrodeoxygenation contain Co/Mo, Ni/Mo, or Ni/Co/Mo, with Co/Mo preferred for being more active.

These high-performance specialty catalysts with designed pore architectures have been used commercially for upgrading heavy crude oils to light deoxygenated low-sulfur and low-nitrogen hydrocarbons. Depending on the physicochemical composition of the heavy oil or residue to be processed, the catalyst particle pore architectural structure can be custom designed for achieving maximum product yield and to reduce the need for costly catalyst regeneration.

To that end, catalyst particles in spherical or extrudate shapes can be prepared by incorporating ground or pulverized combustible biomass, which can include recycled plastics, to produce pores with sizes in the mesoporous and macroporous ranges. A blend of the different-sized combustible materials added before the catalyst particles are shaped, followed by a combustion decarbonization treatment, produces a polymodal interconnecting meso/macro pore/channel architecture. Subsequently, these highly porous particles can be loaded with the selected catalytic metals.

To facilitate processing waste biomass as a combustible carbon pore-forming additive, the raw material is first toasted at 105 to 140 °C. Toasting eases grinding and micronizing the material to 1 μm particles or smaller and requires less energy overall. In the same sense, pretreatment by torrefaction can ease grinding; more on torrefaction later. Ultimately, the amount of carbon additive used to create the desired macroporosity and the number of mesopores and micropores formed must be balanced against the loss of the physical strength and attrition resistance of the catalyst particles, which decreases as the total porosity increases.

Furthermore, the low-cost pore-regulating combustible carbon agents can also be used to prepare hydrocracking catalysts containing ultrastable faujasite zeolites. These wide-pore zeolites, which have been partially dealuminated or desilicated, exhibit a supplementary mesoporous architectural crystal structure and can be additionally phosphated. Homogeneously dispersing the zeolites into a metal-oxide matrix, such as gelled acid peptized pseudoboehmite alumina, or into a silica or silica-alumina gel matrix, along with a small amount of the pore-regulating carbon additives, and further processed as described above, can be used to prepare shaped-catalyst extrudates or spheres exhibiting a macroporous bulk structure uniformly decorated with zeolite crystallites bearing micro- and mesoporous architectures.

We also note that in preparing ultrastabilized zeolite catalysts the final step involves a mild acid washing, which removes some of the alumina moieties thus creating mesopores in the zeolite crystallites. This processing conducted at commercial scale represents a low-cost mesostructuring approach. Other techniques for creating architectural mesoporosity involve using surfactant templates to increase crystal bulk surface area and pore volume, though it will decrease the heat capacity and heat conductivity and decrease overall performance. A low-cost remedy to this problem is to introduce phosphorus. Other low-cost remedies to address the heat-deficiency issue are to include a small amount of red mud or other inexpensive natural material that contain transition metals or rare-earth metals into the catalyst particles [58].

In yet another twist, low-cost catalysts with customized architectural microporosity can be prepared from a silica-alumina co-gel containing combustible microfine biomass particles. Then following microsphere formation the particles can be irradiated with ionizing gamma rays. This approach imparts catalytic activity by boosting the number of active sites, reducing the dependence on rare earth or other catalyst metals and production costs [70].

In addition, the pore-regulating additives prepared from low-cost pulverized biomass to a chosen particle size can be used to create macropores in catalyst supports, prepared according to the Engelhard/BASF process, by incorporating the carbon additive in a kaolin clay slurry before being spray dried to form the support microspheres. Subsequent calcining and burning out the carbon additive leaves macropores in the bulk of the microspheres. Following that, zeolites can be added to the macroporous material via the classic Engelhard technology for producing commercial FCC-type catalysts, where the zeolite phase can be modified by ion exchange with catalytic metals, ultrastabilized, modified to form mesopores, and also phosphated.

Overall, we reiterate that the presence of meso- and macropores in the catalyst particles enhances the diffusion rates of hydrocarbon reactants and product molecules into and out of the particles, which increases catalytic activity and decreases both coke formation and catalyst deactivation rate. Furthermore, similar catalysts also provide an opportunity to convert bioethanol derived from corn, sugar cane, sugar beets, and other crops to light hydrocarbons for transportation fuels and commodity chemical applications.

Returning to preparation of FCC catalysts, when those made according to the Engelhard/BASF technology were tested in a biomass conversion pilot plant, they produced more bio-oil and less coke at the same degree of deoxygenation as compared with a conventionally prepared catalyst—one made by homogeneously mixing ZSM-5 crystals, kaolin clay, and silica binder in a water slurry and then spray-drying the mix to form the microspheres followed by calcining.

Conceptually, one could consider further cost savings to prepare the desired microsphere particles by simply coating sand particles with zeolite crystals. However, attempts at forming composite particles by synthesizing a ZSM-5 from conventional chemical solutions using an organic template under typical process conditions in the presence of sand particles have been unsuccessful. The zeolite-formed crystallites cannot adhere strongly enough on the relatively smooth surface of the sand particles to withstand subsequent catalyst preparation steps or biomass reactor conditions—a fine zeolite powder would form and foul the reactor equipment, plus be difficult to separate from the bio-oil. This failure to attach zeolites on sand also takes place even when the smooth surface of the sand particles is chemically etched with sodium hydroxide to leach some silica from the particle surfaces, creating structural defects for zeolite adhesion.

The idea of crystallizing zeolites on a hard substrate surface goes back more than 20 years, when George R. Gavalas at California Institute of Technology initiated pioneering research on synthesizing zeolitic films and porous membranes on surfaces, such as growing ZSM-5 films on porous α-alumina particles and other porous substrates including silicon carbide [71]. Gavalas developed these composite molecular sieve materials for use in gaseous mixture separations, but we are mentioning here the composite material made from a SiC microspheroidal core coated with a film of ZSM-5 or any other small-pore zeolite, because this composite would be an ideal catalyst and heat conductor/exchanger for thermocatalytic conversion of biomass, given SiC's efficient thermal properties. However, the cost of commercially producing this molecular sieve support matrix composite materials may prohibit their use in biomass-to-liquid fuels conversion.

Furthermore, it is possible to improve zeolite catalyst performance by calcining mixtures of anhydrous and calcined kaolin powders to form dense, high-heat-capacity spinel/mullite phases and by compounding these materials in a water slurry with ZSM-5 crystals and silica binder, followed by spray-drying to form microspheres. This type of catalyst subsequently was tested for biomass conversion as well, which resulted in an improvement with respect to coke formation, bio-oil yield, and deoxygenation as compared with the performance of conventional ZSM-5-based commercial catalysts. However, at KiOR its catalytic performance was overall inferior to the catalysts prepared by the Engelhard/BASF technology and would have cost more.

3.6 CHOICE OF BIOMASS STARTING MATERIAL AND OTHER NUANCES

Meanwhile, not to be ignored in the economic picture of biofuel and biochemical production, is the need to have an economically sustainable, globally available supply of biomass and convertible organic waste materials. That means effective use of whole plants, including harvested grasses and trees, as well as forestry and agricultural wastes that don't require pretreatment, such as removing bark, before conversion to bio-oil. It is imperative that we include among these sources nonfood biomass such as wheat straw and corn stover, along with household, business, and industrial waste paper and cardboard and unrecyclable plastics, which can be mixed in with the primary biomass source in different proportions. One ton of debarked high-quality pine wood, when thermally converted under optimized conditions in a CFB plant using sand as the heat-transfer medium, will produce about 80 to 90 gallons of crude bio-oil—which was KiOR's production target. However, when the whole tree is processed and similarly converted, it produces in the same pilot plant only about half the volume of bio-oil compared with debarked pine.

As mentioned at several points earlier, an inherent challenge is that some low-grade, low-cost biomass typically contains significant amounts of undesirable indigenous metals, in some cases up to 25% by weight of the dry material. When thermally devolatilized in a CFB reactor using a noncatalytic heat-transfer medium such as sand, biomass containing high metal content produces relatively low bio-oil yields, partly due to the accumulation of catalytically active ash. The ash, made up mostly of metal oxides, promotes side reactions, which mainly includes gasification of the nascent formed bio-oil along with formation of char. As an aside, indigenous metals provide an advantage if gasification to achieve combustible gases is the goal. In a one-reactor system, the metals/metallic ash interacts with zeolite active sites, exchanges with the zeolite's native metals or replaces the zeolite's hydrogen cations, and blocks catalyst pores, for example, when being dissolved in the water phase as metal salts, thereby deactivating the catalyst and requiring higher rates of catalyst regeneration and replacement—with added cost.

The present challenge to be solved is that low-grade, low-cost biomass feedstocks are needed, because they are abundant and readily available worldwide, but they contain a troublesome amount of metals. Yet, the only catalysts known as being active and selective for biomass deoxygenation are the small-pore ZSM-5 zeolites, which it bears repeating are costly, hydrothermally unstable, and easily deactivated by the acid water/vapor and metals from the biomass.

From this discussion, it becomes clear that removing the intractable biomass components, including metals, increases liquid hydrocarbon yields, with evidence for doing so now growing [72]. These components can be removed to different degrees by treating the raw biomass with solvents, acids, and bases [73-74].

However, most of the published pilot-plant data since the original pioneering work at Occidental Petroleum and at the University of Waterloo in the early 1980s [13,17-19] involves using high-quality lignocellulosic biomass, for example debarked pine chips or sawdust with low metal content. The development and use of efficient robust catalysts and suitable processes to convert low-quality, high-metals biomass resources, which are essential for a global cost-effective, economically sustainable, and profitable biofuels business competitive with petroleum, have essentially been neglected.

In scouting for a solution, technologies for deashing raw biomass started to be developed in the late 1960s, beginning with the pioneering work by Fred Shafizadeh at the University of Montana [75-76]. Subsequently, Caltech's Gavalas and coworkers carried out innovative work on scalable biomass demineralization, showing that treatment with acidic water rinses is sufficient to remove metals, but not severe enough to cause hydrolysis of the woody biomass [77]. The demineralization leads to substantially increased organic liquid yields during biomass conversion. These findings have since been repeated and validated by other research groups [78].

By 2010, researchers had recognized that most of the metals must be removed from the biomass raw material before conversion to fuels. However, when taking into consideration the additional cost and environmental impacts of demineralization at commercial-scale production, including procuring the acid used, treating/disposing of large volumes of wastewater, and generating heat for drying the wet biomass that may contain up to 80% by weight water, then the overall costs, once again, render these processes economically unfeasible, as confirmed commercially by KiOR.

Seeking improvements, Stijn Oudenhoven and coworkers at the University of Twente have described an interesting and useful technique for a more cost-effective removal of indigenous metals by reusing the acidic water formed during biomass thermolysis as a solvent in a demineralization pretreatment step of raw biomass before it is devolatilized [79]. Furthermore, a team at KiOR developed an effective deep demineralization process by using recycled acidic hot water in conjunction with mechanical treatment [80]. The KiOR approach takes advantage of the natural tendency of the biomass to swell by absorbing water. The mechanical treatment squeezes out the absorbed water containing dissolved metal salts [81]. Conducting this process at elevated temperature shortens the treatment time, and the process can be repeated to achieve a higher degree of demineralization.

The advantages of these biomass demineralization processes are the use of relatively small volumes of water or acidic solutions through recycling, shorter processing times, and more effective metal removal, as compared with simple static biomass washings with water or acidic solutions. A more technically feasible and cost-effective demineralization process would be a combination of two technologies—the Oudenhoven approach and the KiOR approach—using/recycling the acidic water formed during biomass thermolysis and applying mechanical action to the biomass feed, with optional heating.

One shortcoming of demineralization pretreatment of the raw biomass is that removing the metals and increasing the porosity of the biomass particles reduces the material's volumetric heat capacity. Consequently, it is imperative to include a suitable high-efficiency heat-transfer medium in the conversion reactor as discussed previously for quickly heating the biomass to an optimal devolatilization temperature. This rapid, preferably "flash" heating still must be balanced by keeping the regenerator unit temperature as low as possible, to avoid damaging the regenerator equipment and the catalyst, also keeping in mind that more heat means more cost.

Besides demineralization, another option to reduce bulk biomass metal content in the reactor mixing zone is to develop a hybrid production system in which high-quality biomass with low metal content is blended with a portion of inexpensive low-grade biomass feedstock containing higher amounts of metals. This approach is attractive to help reduce cost and improve bio-oil yield when feedstock availability is limited.

In the same vein of using mineral waste for catalysts and recycled acidic water for demineralization, the waste noncondensable gas by-products of biomass devolatilization, that is CO, CO_2, and CH_4, can be put to good use. In particular, carbon monoxide with its strong chemical reducing properties can be used to deoxygenate the nascent produced bio-oil and thereby reduce the amount of catalyst needed, as well as reduce the volume or even replace the costly hydrogen used in the subsequent bio-oil deoxygenation or other upgrading processes.

These gases can be recycled to the reactor and brought into contact with the raw biomass feed and catalyst/heat-transfer material while mixed with the regular nitrogen gas lift in the CFB reactor. This by-product gas-recycling treatment as a means of reducing bio-oil oxygen content in a CFB biomass conversion reactor has been described in a KiOR patent [82]. Charles A. Mullen and colleagues have further published an enlightening study of recycling the generated noncondensable gases into the reactor during biomass devolatilization [83].

Another cost-effective means of reducing bio-oil oxygen at the front end of the overall conversion process, as hinted at earlier, involves using waste plastic from manufacturing processes or recycled household plastics and other organic wastes to supplement the biomass feed. Besides having low or no oxygen content, the plastics provide additional hydrogen for use to remove oxygen from the oxygenated hydrocarbons produced from the raw biomass, as described in a KiOR patent [84]. This approach, especially if combined with using wastepaper/cardboard and organic solid wastes, has the additional benefit of lowering the burden on landfills and recycling costs, as well as reducing air pollution and greenhouse gas emissions from landfills and municipal solid-waste incineration in conjunction with international efforts to preserve atmospheric ozone, reduce the rate of global warming, and minimize destructive climate change.

The use of certain bifunctional catalysts can further reduce the overall cost of crude bio-oil upgrading by providing hydrogen to the reaction system, which can help reduce oxygen in the bio-oil via hydrothermolysis. Ultrastable low-sodium ammonium wide-pore faujasite and small-pore pentasil silicalite molecular sieves/zeolites can be produced in large commercial quantities at relatively low cost by using an ammonium salt in a continuous ion-exchange process. When the ammonium zeolite particles are introduced into the catalyst regenerator unit operating at temperatures in the range of 700-760 °C, the ammonium ions thermally decompose, producing ammonia and hydrogen ions, with the ammonia decomposing to N_2 and H_2, which is needed for hydrotreating. The acidic hydrogen ions go on to react with zeolite crystal lattice oxygens, breaking oxygen-metal lattice bonds and forming hydroxyl groups that further thermally dehydroxylate and cause lattice reconstitution leading to ultrastable zeolites, known as USY types, similar to acidic dealuminations with intermediate calcinations.

An even more effective hydrothermolytic system could be created by using the ammonium ion-exchanged zeolites as hydrogen ion donors for processing blends of biomass and waste plastic as noted before, producing lower oxygen content bio-oils and light hydrocarbons with greater yield. The intermediate products subsequently will require less severe reaction conditions, less hydrogen feed, and lower amounts of hydrocracking and hydrotreating catalysts for producing biofuels and specialty chemicals.

The overall biomass/waste biomass to energy and specialty chemicals conversion process described here provides a technoeconomically feasible means to additionally manage the in-process generated CO_2, and CH_4, which are the main components of global greenhouse gas emissions. These gases along with the char generated in the biomass devolatilization reactor or during crude bio-oil upgrading can be directed to the heater of the regenerator for combustion. The hot gases produced there, that is residual CO_2 and water steam, can be circulated in a thermally insulated pipe system to provide heat in other processing steps, such as drying the biomass feed. Finally, this gaseous mixture can be separated, and the leftover CO_2 sequestered and stored and/or utilized as a C1 chemical feedstock. This separation can be carried out using one of the known commercially used processes, such as pressure swing adsorption utilizing zeolites as sorbents when the concentration of the CO_2 in the mixture is high; alternatively, cryogenic, solvent, or membrane types of separations can be used.

SIMPLE BUT ESSENTIAL HYDROGEN

Hydrogen is the simplest element in the universe and hydrogen gas is the simplest molecule. H_2 is also one of the most essential components and process cost variables in petroleum refining, as well as for treating metals, producing fertilizer, and processing foods (converting unsaturated fats to saturated fats and oils). During refining, significant amounts of H_2 are required to upgrade crude oil via hydrocracking and hydrotreating, and in a biorefinery H_2 is more critically needed to remove oxygen when producing light hydrocarbons as useful fuel ingredients. The most common way to produce H_2 is steam reforming, a process in which natural gas (largely CH_4) is combined with steam at about 850 °C and catalytically cracked to form H_2, CO, and CO_2, a mixture known as synthesis gas or syngas, which is also used to make ammonia and methanol; CO is converted into CO_2 with steam in a further step during which more H_2 is produced. The H_2-rich product is subsequently purified to the desired quality level.

To that end, several ongoing R&D efforts to utilize waste gases have been reported, involving companies as varied as ExxonMobil, Linde, and Mitsubishi Heavy Industries, with some advancing to commercial production. For example, Blue Planet uses its Liquid Condensed Phase technology to convert captured CO_2 into synthetic limestone for use in building and highway materials such as bricks and concrete. In addition, bioenergy production with carbon capture and storage is a new approach, known by the acronym BECCS. On this front, engineering and construction project management firm Bechtel has entered into a partnership with renewable energy company Drax to identify opportunities to construct BECCS power plants. In another example, in February 2021 ExxonMobil created a venture called Low Carbon Solutions, effectively a start-up business run within the giant oil company, to cut emissions through carbon capture and underground storage. The initiative will take CO_2 from various sites, including from other companies, and pump it into wells on sites ExxonMobil owns in Louisiana. ExxonMobil figures to turn a profit. On still another front, green methanol derived from CO_2 produced in cement plants and steel plants is increasingly attractive, more as a feedstock replacement for coal in making olefins to produce plastics than it is for green energy uses, with companies such as Carbon Recycling International and Mitsubishi Gas Chemical getting in on the action. Still one more intriguing example is Air Company, which hydrogenates CO_2 to make ethanol, that is its "Air Vodka." Air Company is also aiming to produce sustainable aviation fuel and announced in 2022 that it has buyers for its first one billion gallons, including the U.S. military and commercial airlines. These advance purchase agreements are common, as noted earlier for KiOR. But they don't count until the product is delivered, and so far, Air Company is still at the pilot-plant stage. Overall, some three dozen negative emissions technologies for carbon utilization and storage have been proposed, ranging from direct air capture and BECCS, to generating and burying biochar, to enhancing ocean alkalinity to boost carbonate formation.

Still another gadget in the chemical engineering toolbox to optimize commercial refinery processes is heat management. In particular, process engineers must take into consideration how to use excess heat produced by the catalyst regenerator and how combusting the unwanted char to produce heat can be used to an advantage—for example, redirecting the heat where it is best used, such as drying and preheating the raw biomass.

In treating raw biomass, heating in the range of 200-300 °C is a process called torrefaction [85], whereas milder treatment in the range of 105-200 °C is referred to as toasting [69]. The benefits of heating include less weight when the biomass is to be transported and more brittle biomass particles that are easier to grind to uniform size and are hydrophobic. Both torrefaction and toasting devolatilize the raw biomass to different extents while producing a material with higher energy density and higher heat capacity and conductivity. The subsequent thermolysis processing thus enjoys more homogeneously fluidized biomass particles that better mix with the catalyst and/or heat-transfer medium in the CFB reactor. Furthermore, the heat-pretreated biomass enters the CFB reactor containing much less water, meaning less heat is needed for vaporization and less acidic water is produced. For the latter benefit, this means less zeolite catalyst destruction and less water to separate from the bio-oil.

Both the torrefied and toasted biomass feeds lose CO, CO_2, and light hydrocarbons to varying degrees during their thermoprocessing; hence, cost-wise, the materials advantageously contain less water and less oxygen to remove during subsequent upgrading steps to produce transportation fuels. However, this gain needs to balance against carbon value lost. Some lost carbon value can be recovered if, as noted earlier, these gases are recycled to the biomass devolatilization reactor to enhance deoxygenation of the crude bio-oil.

"Flash heat drying" is another kind of raw biomass thermal pretreatment, which can be useful when the biomass contains a lot of water. For example, pine chips contain close to 60% by weight water. With commercial flash dryers, rapid heating of the water sorbed in

the biomass particle bulk matrix structure causes sudden vaporization, generating pressurized steam that can disrupt and open up the
densely packed cellulose, hemicellulose, and lignin components. The
flash-drying pretreatment, which resembles on a much smaller scale
the well-established "steam explosion" pretreatment of raw biomass,
results in an increased yield of bio-oil in subsequent devolatilization.
As a bonus, noted previously, the excess heat produced by the catalyst
regenerator, or by burning waste char, can be funneled in to run the
flash dryer.

The additional advantage of both torrefaction and toasting relates
to the hydrophobic nature of the biomass particles—reducing the
amount of water present makes the raw biomass more lipophilic and
thereby more compatible with oil phases and amenable to forming
uniform low-solids suspensions. This property allows for coprocessing
bio-oil with oil from petroleum refinery streams, such as hydrotreated
vacuum gas oil, which is the portion of crude oil used to make gasoline. The bonus here in coprocessing is that the hydrotreated refinery
oil stream donates hydrogen to the hydrogen-poor bio-oil phase for its
deoxygenation. Alternatively, the bio-oil can be produced in a separate
CFB rector using sand as the heat-transfer medium, and part of the
produced bio-oil vapor or the bio-oil can be blended with a refinery oil stream and thermocracked in an FCC unit to produce light,
deoxygenated hydrocarbons. Most important, the heat-pretreated
fine-particle biomass, preferably prepared in a flash-heat dryer, can
be homogeneously dispersed in bio-oil and the slurry formed can
be thermally treated in an autoclave at an elevated temperature and
pressure in a hydrogen atmosphere to convert the dispersed biomass
particles into bio-oil [86]. In total, these biomass heat-treatment steps
provide another useful path to increase the efficiency and reduce the
production cost of the thermal and chemical conversion of biomass to
clean fuels and chemicals.

Taken together, procuring and using the catalysts, sourcing raw
biomass (including transportation costs), and generating upgrading
hydrogen gas (which includes overall process heat management) are

the three major cost-driven variables in the commercial production of biofuels and biobased chemicals. Their total expenditure is what determines the make-or-break economics of a large-scale renewable fuels and chemicals business. As the business slogan goes, it has to be "Big, Clean, & Cheap." The question is: How do we get there?

4.0 A GOLDILOCKS APPROACH TO SUCCESSFUL BIOMASS CONVERSION

When lower SAR crystalline small-pore silicalite zeolites, such as ZSM-5s, are exposed to the aforementioned acidic hydrothermal conditions, they exhibit weak resistance and incur compositional lattice changes, as has been duly noted. As part of these complications, the changes include dealumination to produce nonframework alumina (NFA) chemical species [87]. The hydrophilic nature of the low SAR ZSM-5s enhances these processes. But as the SAR increases, the zeolite crystallites become less hydrophilic, or rather more hydrophobic, which contributes to increasing the stability against the destructive hot acidic conditions in the reactor.

The result of NFA formation, which can be described as a self-reconstituted "partially aluminum-purged crystal lattice," is a higher framework SAR with fewer but more stable catalytically active sites. Most of the NFA species migrate to the exterior surfaces, to pore surfaces, and to channel surfaces of the zeolite crystallites and remain there in chemically reactive forms. As such, they are available to interact with the biomass indigenous metals to form mixed-metal aluminum oxides that set up roadblocks for bio-oil hydrocarbon molecules by plugging zeolite pores and restricting diffusion in and out of the zeolite particles.

By impeding intracrystalline and in-and-out diffusion, and

thereby increasing the residence time of reactant and product hydrocarbon molecules within the pores and channels, molecular and free-radical side interactions are enhanced. The result is increased formation of coke, mostly in the micropores, which reduces substantially the activity/selectivity/deoxygenating ability of the catalyst. This effect has been demonstrated in biomass conversion to bio-oil in pilot plants and in a semiworks plant, using a low SAR ZSM-5 catalyst with continuous catalyst regeneration on-stream for long runs. Additionally, the newly formed mixed-metal aluminum oxides derived from NFA species act as catalysts, promoting gasification of the bio-oil, which also reduces bio-oil yield. This effect is more pronounced and costly when low-quality biomass feedstocks with high metal content are used.

Low SAR HZSM-5 zeolites are even less hydrothermally stable. The hydrogen ions interact with and hydrolyze the weaker bonded alumina-oxygen-silica lattice sites, forming silica-hydroxyl groups and crystal lattice defects during high-temperature treatment in the catalyst regenerator unit, leading to dihydroxylation, dealumination, and NFA formation. These effects, practically eliminating nascent catalytically active sites of the zeolite, are especially a problem when low SAR HZSM-5s, such as the widely used SAR 24 commercial catalysts, are on-stream for long operating periods in single-reactor systems with continuous catalyst regeneration. It is re-emphasized here that the severe conditions to which the zeolite catalyst is repeatedly exposed is the reason it is strongly recommended to remove the acidic water/vapor phase formed in the first reactor before the bio-oil is introduced in the second reactor where the zeolite catalyst is used.

Note that coking deactivation of a catalyst can be corrected during catalyst regeneration, whereas crystal lattice dealumination, reconstitution, and catalyst active-site elimination are typically not repairable: It is imperative to minimize these effects early in the catalyst lifetime. Furthermore, higher SAR ZSM-5 zeolites and their ion-exchanged forms with elements such as calcium, magnesium, gallium, iron, or zinc are more stable toward dealumination and exhibit lower rates of deactivation when on-stream in long runs, compared with their HZSM-5 analogues. Also of interest is that larger pore size materials, such as mesoporous MCM-41, prepared from ZSM precursors, which we propose above to produce at low cost, share similar generic hydrothermal properties with the microporous pentasil silicalite-type ZSM-5s. As mentioned before, higher SAR MCM-41 materials are thermally more stable and produce less gases/coke and more bio-oil than the less stable low SAR grades, as demonstrated in pilot-plant studies [63-64,88]. Furthermore, synthetic wide-pore faujasites and their ultrastable forms, most of them ion-exchanged with rare-earth metals, are still predominantly used in cracking and hydrocracking refinery units globally.

These zeolite and catalyst concepts have a rich history, reaching back more than half a century to when synthetic zeolites started to be commercialized and used as catalysts in petroleum refining. Stamires, a coauthor of this account, participated in their original discovery and development during the commercialization of synthetic molecular sieve faujasite-type zeolite catalysts for petroleum cracking and hydrocracking at Union Carbide starting in the late 1950s [89-92]. Later on, Stamires worked as a consultant and then as Vice President of Research & Development for Filtrol/Kaiser Aluminum Corp., followed by similar roles at AkzoNobel and Albemarle. At Akzo, he introduced new high-performance zeolites and cracking and hydrotreating catalysts that enabled the company to expand its catalyst business with advanced technologies and build new catalyst production and regeneration plants in the U.S., Brazil, Japan, and the Netherlands to become a major global catalyst supplier.

4.1 OPTIMIZING CATALYST SELECTION

KiOR's work indicated that higher crystallographic SAR small-pore pentasil silicalite zeolites should be used for cost-effective commercial biomass conversion operations. To achieve optimal long-term catalytic performance and cost effectiveness, these ZSM-5s should have SARs above 25, preferably in the range of 35 to 45, but not above 50. This is because the catalytic activity and selectivity decrease with fewer acidic aluminum active sites present in the zeolite crystallites, resulting in lower catalytic deoxygenation activity.

It is worth repeating that the lower SAR zeolites exhibit a misleading higher initial fresh activity, sometimes referred to as "overcracking," which is short-lived and causes excessive early coking of the catalyst; however, these catalysts with repeated regeneration incur lattice self-dealumination and reconstitution-ultrastabilization to equilibrate quickly to a higher SAR with lower but more stable activity levels. These targeted higher SAR values do in general ensure sufficient catalytically active sites and catalyst stability to maximize the lifetime and selectivity of the catalyst on-stream, requiring less catalyst and less frequent spent catalyst replacement, in the same fashion as that for the wide-pore faujasite-type zeolites used globally in FCC and hydrocracking catalyst formulations [93].

A key point in this discussion is that, for cost-effective commercial operations, the optimum ZSM-5 SAR needs to be experimentally determined for a given process condition, such as biomass feedstock type, different types of biomass process conversion, and product yield optimization (Figure 10).

That is not the whole story, though. For realistic evaluation and selection of a catalyst for long-term steady performance, the catalyst is usually partially steam deactivated and regenerated to equilibrate the activity and to reduce overcracking and overcoking before being introduced into a reactor; a more detailed discussion with data on coking and the detrimental effect on bio-oil yield is available [94]. As outlined earlier, NFA moieties form and thus block pore openings and screen catalytic sites during this high-temperature treatment step. Therefore, a mild acid washing of the steamed catalyst followed by drying can remove some of the undesirable NFA moieties and improve performance.

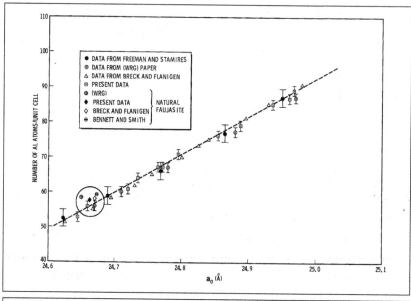

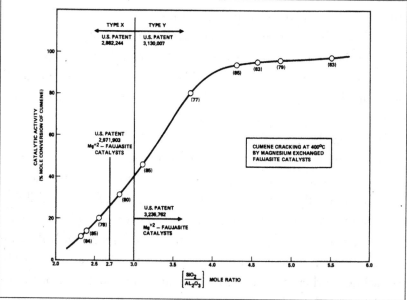

Figure 10. Comparison of the linear relationship between crystal unit cell chemical composition (top), in this case aluminum atoms in the cell and unit cell size, for a set of synthetic and natural faujasites [93], and the linear variation of catalytic activity of faujasite catalysts (bottom) with different crystal lattice compositions (Si/Al mole ratio, or SAR).

In summary, this partial, mild dealumination of ZSM-5s increases the intercrystalline diffusivity of reactant and product molecules in the crystallites, unblocks the screened active sites, and decreases the residence times of reactive molecules, intermediates, and free radicals within the zeolite crystallites, resulting in the formation of smaller amounts of coke and higher product yields. In addition, this mild acidic dealumination partially modifies the inherent zeolite parent microporous crystal structure by increasing its lattice SAR and introducing into the zeolite crystallites a supplementary architectural mesoporosity, which enhances the efficiency to increase bio-oil yields and decrease amounts of by-products [95].

In providing a bit more historical background, the larger pore faujasite-type zeolites synthesized in a sodium cationic form (zeolite Y) were invented by Breck and developed commercially at Union Carbide [35]. Commercial grades for zeolite Y have SARs from 3.6 to 5.5, whereas commercial grades of zeolite X type have SARs from 2.3 to 3.6. Subsequent research and development at Union Carbide's Linde Division involved treating molecular sieve sodium faujasite zeolites with SARs of 4.0 or greater with repeated ammonium ion exchanges with intermediate calcinations and mild acidic dealuminations for removing the formed NFAs. The resulting new highly stabilized zeolite product materials were originally described by Jule A. Rabo and colleagues at Union Carbide, including Stamires, Paul E. Pickert, and James E Boyle. They were denoted as decationized molecular sieves [36,96] and subsequently referred to as ultrastable Y, or USY, zeolites. They have been used globally in preparing hydrocarbon conversion FCC and hydrocracking catalysts. Besides having an increased SAR, they also have a partial supplementary architectural mesoporosity with increased hydrothermal stability, with the chemical mechanism of decationization, ultrastabilization, and USY zeolite formation originally discovered by Herman Szymanski, Stamires, and colleagues [89] and later rediscovered by C. V. McDaniel and P. K. Maher [90]. They are presently used commercially as hydrocracking catalysts for upgrading heavy oils, including bio-oils, to light hydrocarbons [97-98].

The sequential catalyst discoveries at Union Carbide practically revolutionized the global petroleum refining industry: First the discovery of synthetic molecular sieve faujasite type Y by Breck followed by the discoveries of the ion-exchanged faujasite and the ultrastable faujasite zeolites by Rabo, Stamires, and colleagues, which since that time are mostly used in oil refineries in cracking and hydrocracking processes. Originally, these synthetic faujasite zeolites in hydrogen or magnesium ionic form were used as FCC catalyst materials to increase substantially the conversion of crude oil to gasoline, diesel, and jet fuel products. These developments helped increase refinery/oil company profitability, plus for consumers provided a societal benefit of having more fuels available at filling stations at lower cost and with lower environmental impact. In addition, Mobil Oil's R&D group made a substantial contribution to petroleum processing technology by introducing new FCC catalysts made with rare-earth ion-exchanged synthetic faujasite zeolites [99]. These chemistry developments at Union Carbide and at Mobil Oil were later considered as possible candidates for the Nobel Prize.

Furthermore, since desilication of zeolite crystals has also been used as a procedure to create architectural mesoporosity, it is worth briefly comparing this process with the classical controllable acidic zeolite fractional dealumination procedure used for ultrastabilization and the creation of crystal architectural mesoporosity. This is to say, going in the opposite direction by treating the inherent, as-synthesized zeolite crystal with an alkaline solution to leach out silica and thus producing lower SAR zeolites with mesoporous lattice architecture, though at the cost of reduced hydrothermal stability.

Regarding the change in stability by desilication, the lower SAR zeolites have a larger number of energetically weaker lattice-bonded alumina sites in the mesoporous crystalline framework, and therefore they are less hydrothermally stable and exhibit higher rates of catalytic deactivation while on-stream in long process runs. In addition, as the catalysts undergo regeneration treatments, their zeolite framework self-dealuminates, altering the SAR by producing more NFA moieties, which decreases the catalytic activity and selectivity.

Consequently, for overall cost-effective purposes and process efficiency in commercial operations, a mild acidic fractional dealumination to produce customized zeolite mesoporosity is preferable over the desilication method for the large-pore faujasites and also for the small-pore silicalite pentasil zeolites such as the ZSMs.

Given the prime interest to produce low-cost catalysts to convert waste biomass to commodity fuels and specialty chemicals, when at commercial scale a continuous processing equipment assembly that includes several sequential steps is more cost effective than conventional multi-step batch processing [100]. Also useful for commercial cost-effectiveness is ion-exchange of the nascent, as-synthesized zeolite's sodium ions with ammonium ions in a continuous process to achieve a high degree of ion exchange and produce zeolites containing about 1% sodium (as Na_2O) or less without negatively affecting the crystallographic structure.

Deep hydrothermal ion exchange of zeolites conducted in a single batch reactor has been described [92]. The cost benefit afforded by low-sodium zeolites, which in a continuous process can include ion exchange with rare-earth metal ions, is that the process can be incorporated with forming the catalyst particles from a slurry with a binder and an inert matrix material added and then spray dried or extruded, with the formed particles subsequently calcined. These compounded low-sodium zeolites, in addition to their microporosity, form a supplementary mesoporosity as mentioned earlier, which together with the absence of sodium exhibit increased hydrothermal stability, activity, and selectivity for producing gasoline with increased octane number.

On a separate note, these catalyst particles can be directly introduced into the catalyst regenerator unit of an oil refinery plant operating at high temperature and cause an in-situ crystal lattice self-ultrastabilization of the zeolite component. Still further, the low-sodium catalyst particles containing predominantly ammonium ions can be calcined before use, and also be subjected to a mild acidic solution treatment, to remove the formed NFAs and create a supplementary mesoporosity of the zeolite crystallites present in the catalyst particles. For commercial large-scale operations, this process is simpler and more cost-effective as compared with other known zeolite ultrastabilization batch processes involving several ion-exchange/dealumination steps with intermediate calcination and/or steaming steps [101-103].

Because of the growing interest in today's literature discussing the preparation and applications of synthetic zeolites for bio-oil production and upgrading to fuels and specialty chemicals, with most reports involving changes in zeolite SAR, a clear distinction needs to be made about the use of the term "SAR." The term nowadays is being used rather loosely without specifying what it really represents, which should be the crystallographic elemental composition of the unit cell as determined by X-ray diffraction, and not the bulk chemical composition of the commercial zeolite product, as determined usually by standard chemical elemental analysis.

Crystallographic SAR determination from X-ray diffraction patterns of zeolite powders has been successfully used on commercial products to measure the amounts of zeolite formed during catalyst preparation (Figure 11). For example, some years ago, as described in Engelhard/BASF patents, Walter L. Haden Jr and Frank J. Dzierzanowski used the linear correlation of unit cell dimension change with element composition to distinguish between different members of a continuous series of faujasites with varying SARs, not making a distinction between zeolite types X and Y [104]. Stamires and Donald Freeman Jr. developed and used this experimentally derived continuous correlation earlier for the same purpose in confirming the absence of any SAR discontinuation, specifically at SAR 3.6 [93,105]. It should be noted that in producing commercial amounts of these wide-pore synthetic faujasite zeolites, although one adjusts the amounts of individual chemical ingredients added to the reaction vessel to produce a zeolite having a specific SAR, the crystallized product ends up having a uniform distribution of zeolite crystals with SAR values below and above the synthesis target. In general, as the smaller silicon atoms are replaced with the larger aluminum atoms within the unit cell, the unit cell size/volume increases and X-ray diffraction patterns can identify the changes and relate to the

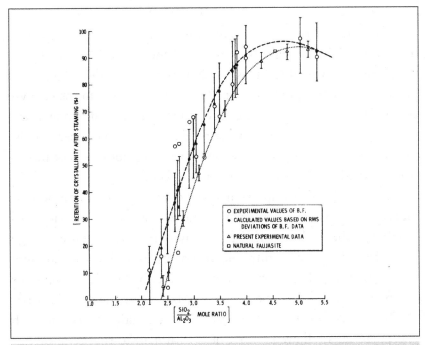

Figure 11. The hydrothermal stability of the zeolite crystal lattice increases with increasing silica content, or higher SAR.

corresponding SAR (SiO_2/Al_2O_3 ratios); alternatively, researchers can use Rietveld refinement procedures.

As an aside, the usual simple pretreatments such as mild steam deactivation or a calcination step of the fresh catalyst before use have a side beneficial effect of causing some densification of the catalyst microsphere particles because of the loss of some bulk surface area and pore volume. Such pretreatments increase particle volumetric heat capacity and heat conductivity and enhance the efficiency of the biomass devolatilization process.

Spent FCC-type catalysts, known as equilibrium catalysts, or ECATs, being denser with lower surface area and pore volume than the predecessor fresh catalysts, also exhibit this increase in volumetric heat capacity and heat conductivity. As expected, these heat property improvements are also realized when the ECAT contains substantial amounts of metals, which are deposited on the catalyst from process feeds, which could be crude petroleum, shale and tar sand oils/bitumen, or bio-oil, although the presence of the metals can have a negative effect on bio-oil yield, as noted before. To an advantage, small portions of ECAT can be drawn from the feeder line to the catalyst regenerator and treated in a weak acid slurry at ambient temperature, ideally using recycled acidic water produced in biomass thermolysis, to partially remove NFAs and some of the indigenous biomass metals before being reintroduced to the catalyst feed.

Another notable improvement of overall thermocatalytic biomass processing performance with some cost savings benefits involves modifying the zeolite in ZSM-5 catalysts by various ion-exchange treatments designed to introduce elements other than metals into the zeolite framework. For example, as described earlier phosphorus can be introduced into the zeolite crystallites or the zeolitic particle shell during microsphere preparation by solution treatment with phosphate salts followed with a heat treatment to fix the phosphorus into the zeolite (incipient wet impregnation techniques). These phosphorus-modified zeolites, the same as zeolites containing alkaline-earth metals, transition metals, or rare-earth metals or various combinations of these metals, can be subsequently processed using standard techniques to prepare catalyst microsphere particles. One example describes an interesting combination of lanthanum exchanged into a ZSM-5 with phosphorous on the zeolite, providing synergistically interacting lanthanum-phosphorous moieties, resulting in improved catalytic performance for producing light olefins from naphtha feeds [106].

As described before, the presence of phosphorus in the ZSM-5 containing microspheres increases the hydrothermal stability of the catalyst and reduces coke formation to improve bio-oil yield. Relating to that effect, it is of interest to point out that the nascent formed reactive NFAs interact with the phosphorus to form aluminum phosphate moieties. This chemical transition helps prevent or reduce the migration and agglomeration of NFAs to a certain extent, thereby

WHAT IS A CLAY?

Clay minerals are important aluminosilicate constituents of sedimentary deposits in Earth's crust that develop plasticity when wet but become hard upon drying or firing. They have been useful to humans since ancient times in agriculture, manufacturing, and medicine—for example in pottery, as pigments, and as antacids. Clays are classified into several groupings with variations in chemical composition, such as kaolin and smectite, with some familiar types including montmorillonite, bentonite, China clay, Fuller's earth, and talc. Talc is a magnesium silicate (general formula $Mg_3Si_4O_{10}(OH)_2$), and KiOR used hydrotalcite, a magnesium aluminate (general formula $Mg_6Al_2CO_3(OH)_{16} \cdot 4H_2O$) that resembles talc, as the original catalyst in its biomass catalytic cracking (BCC) technology.

avoiding the formation and growth of large alumina species. These larger alumina species can physically block the pores and channels of zeolites and hence impede mass transfer of reactants and converted hydrocarbon molecules in and out of the crystallites, which increases hydrocarbon residence time on the zeolite, leading to further increased coke formation.

Of additional note, these formed aluminum phosphate species introduce fire-retardant properties to the catalyst particles, which in turn may enhance heating rates of the catalyst particles and enhance the heat-transfer rates from the catalyst particles to the biomass particles, improving biomass devolatilization efficiency. This use of phosphorus to improve hydrothermal stability and catalytic properties, including selectivity and coke reduction, is described in several patents [107-109].

Furthermore, besides phosphating the zeolite crystallites, the matrix of the catalyst particles containing the zeolite or the materials such as clay that go into making the matrix can be phosphated, providing additional catalytic, physical, and mechanical benefits [110-113]. Accordingly, with an eye toward cost-effective commercial catalyst compositions, KiOR staff proposed a new process comprised of calcining a clay such as kaolin to form a meta-kaolin phase and subsequently using phosphoric acid to partially acid leach the calcined clay in a water slurry under controlled conditions to cause some clay dealumination. A small-pore zeolite can be added to a slurry of aluminum phosphate material with the dealuminated clay together with a silica binder and then spray-dried to form microspheres. The zeolite can be exchanged with catalytic metal cations and optionally phosphated before incorporating it into the matrix. Alternatively, prior to compounding, the zeolite can be modified by leaching out alumina or silica with an acid or a base, to develop a hierarchical supplementary mesoporous crystal structure [114].

In a similar cost-effective process, microspheres made from calcined kaolin/mullite clay can be processed to form ZSM-5 catalysts [67], with the microspheres subsequently treated with a solution containing a phosphorus compound and then dried and calcined. Still further, delaminated clay can be used to form the microspheres, which subsequently are treated with a zeolite and then phosphated and calcined, resulting in phosphorus distributed in both the zeolite component and on the particle matrix. A general synthesis of crystalline zeolitic molecular sieves from silica and alumina sources, derived from post-calcined, acid-extracted kaolin clay, is provided in detail in a 1968 patent by Peter A. Howell, which was assigned to Union Carbide [115].

4.2 "JUST RIGHT" OPTIONS

To add everything up, a preferred option for a cost-efficient and environmentally acceptable biomass conversion pathway is to use a two-reactor process system with optimized catalyst/heat-management materials in which the bio-oil produced in the first reactor is upgraded to a partially deoxygenated light hydrocarbon bio-oil in the second reactor (Figure 12). This "light" bio-oil can serve as a suitable feed for use in subsequent low-cost refining using a minimal amount of hydrogen and inexpensive hydrotreating catalysts.

Additional commercial plant hardware cost savings can be achieved by using existing FCC oil refinery units after having their mixing/reactor chambers and risers modified to achieve ultrafast reactant mixing and large flux heat transfer. With this approach, the adverse effects on the integrity and performance of the catalyst—especially zeolite-containing catalysts—in an in situ one-reactor system caused by the hot and steamy acidic "pressure cooker" conditions, the ash and char buildup, biomass indigenous metals, and thermal abuse during catalyst regeneration are avoided, or at least minimized, as has been described in KiOR patents [41,45-46] and elsewhere [116-118]. As an alternative, if the preference is to enhance biomass conversion, the hot and steamy acidic conditions can be used to an advantage, provided the zeolite-containing catalysts are replaced with sand or preferably with a mixed-metal oxide spinel-like refractory catalyst, being more resistant to the harsh environment, as discussed earlier.

The big picture provided by exploring the above-described technologies in detail is how we can use them effectively to create a technoeconomically successful renewable biobased business to meet the objective of producing clean, environmentally acceptable, and sustainable transportation fuels and chemicals at low cost on a global commercial scale that can compete with or replace their fossil-derived counterparts. We can say that we do have available an array of viable, enabling, mutually complementary tools to choose from: These include locally sourced low-metals/high-grade biomass derived from agricultural, forest, and organic municipal wastes to blend with inexpensive high-metals/low-grade biomass and waste plastics to minimize overall processing costs when using dual-functional low-cost heat-transfer materials/catalysts prepared at least in part from natural or by-product minerals that are used in conjunction with process water and reaction gases that are recycled for use within the overall process.

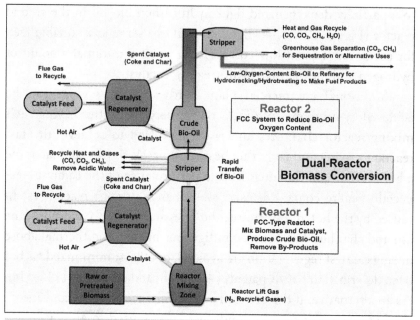

Figure 12. Optimized Dual-Reactor Thermocatalytic Biomass Conversion. In an ideal two-reactor system to maximize crude bio-oil yield, biomass thermolytic devolatilization would take place in the first reactor using a refractory catalyst material or sand to maintain heat. The crude bio-oil produced is rapidly transferred and thermocatalytically upgraded in the second reactor, using an FCC-type catalyst. Quickly removing the bio-oil from contact with noncondensable gases, acidic water vapor, and metal-containing char in the first reactor minimizes side reactions and improves yield. Further processing in the second reactor leads to low-oxygen-content bio-oil suitable for further processing to make fuel products for standalone use or for blending with petroleum or refinery intermediates. This type of reactor was demonstrated on a lab scale at KiOR but not commercially developed. Each reactor would have its own catalyst regenerator, and further have process systems to recycle heat, noncondensable gases, and process water to maximize efficiency, as well as to trap greenhouse gases for alternative uses. The catalysts could be prepared at least in part from natural or by-product minerals/mining wastes, and the biomass feed could be raw or pretreated and optionally supplemented with organic municipal wastes including plastics.

These tools go hand-in-hand with minimizing the number of optimized processing steps for bio-oil separation and recovery, as outlined in KiOR patents [119-120], using large commercial plants to take advantage of economy of scale. We can arrange these tools as necessary to assemble a customized, commercial biomass-to-biofuels/biobased chemicals process and affordably produce the most desirable, in-demand products with a low or negative carbon footprint—just what we need, a "just right" approach. But we must keep in mind that due diligence is needed by conducting reliable preliminary pilot-plant optimization experiments to ensure the viability of available feedstock/feedstock blends, catalysts, and other process variables for making the desired products and optimizing the use of waste and by-products, including greenhouse gases, before going to commercial scale.

5.0 WHAT THE FUTURE HOLDS FOR BIOBASED FUELS AND CHEMICALS

The U.S., Brazil, China, European Union, and other countries and regions around the world have generously spent millions upon millions of dollars supporting R&D projects exploring different pathways to convert biomass into fuels and chemicals efficiently in an economically feasible manner. There have been few successes.

In the U.S., about 10% of gasoline is now displaced by ethanol. Biodiesel and biobased specialty chemicals have made inroads, but still can only claim a small share of the market against petroleum-based products. In Brazil, about 25% of transportation fuel is ethanol and about 20% of all energy is based on ethanol production; the country has taken a lead in using sugar cane, soybean oil, and beef fat as resources to produce sugar, ethanol, biodiesel, and electricity in biobased integrated production facilities developed over the past 40 years [121]. Developments in the rest of the world lag far behind; Iceland is an anomaly, with most of its total energy demand produced from renewable sources—geothermal (65%) and hydroelectric (20%), with little from fossil fuels (15%).

KiOR's extensive experience, from the cradle to the grave, provides a valuable multifaceted lesson to be learned by researchers and investors involved in biomass processing efforts going forward, and indeed to be learned by all of us across the planet who are dependent on the successes. In part, this experience involves how a company such as KiOR communicates transparently and truthfully, both internally and with the public and investors, even when progress is slow or heading in the wrong direction. For example, KiOR's troubles with fundamental technical problems and poor operational results, amplified by management stating exaggerated biofuel yields and markedly deflated production costs, was not disclosed to the Board of Directors and did not start becoming common public knowledge until after six years, in late 2013.

Loescher, KiOR's Vice President of Technology, cemented KiOR's position that the technology was doing well as planned on September 3, 2013, during a presentation at the International Conference on Thermochemical Biomass Conversion Science (tcbiomass 2013) in Chicago. Loescher's talk, "The Path to Commercialization of Drop-in Cellulosic Transportation Fuels," was published in the conference proceedings [122]. It should be noted that Loescher prepared his presentation and submitted his proceedings article without informing most of KiOR's key technical personnel, who learned about it after the conference through acquaintances from outside the company who attended.

Loescher stated that in 2011 KiOR was producing 67 gallons of fuel per dry ton of biomass at a cost of $1.80 per gallon. These numbers had been disclosed in KiOR's SEC application for an IPO of stock. Loescher included that the yield had subsequently improved to 72 gallons, as demonstrated at the Columbus I commercial plant. KiOR's management had used these bio-oil yields and costs in its financial model to estimate profitability. The publicly stated production cost of transportation fuels at $1.80 per gallon did slowly increase to $2.60 and then to more than $3.00, but the announced volume numbers did not change. Recall that KiOR's initial announced long-term goal was to reach 92 gallons per dry ton and become profitable by 2015, and the company had held on to this hope, no matter that it was an unrealistic expectation, and did not want to disappoint investors.

However, the measured production and cost, given the directionally established decreasing trend of bio-oil yield with increasing plant throughput capacity, was actually just 22 gallons per ton and above $6.00 per gallon. The reality hit home at the end of 2013 when Columbus I was supposed to be producing some 13 million gallons of fuel per year but was only able to produce and ship out less than one-tenth that volume to customers who had contractual sales agreements in place.

Most of the bio-oil that was produced was being upgraded to light hydrocarbons for making gasoline and diesel fuel using KiOR's own hydrotreaters and sold to Hunt Oil Company. KiOR also had a once-promising conditional agreement with Catchlight Energy, a Weyerhaeuser-Chevron joint venture. Weyerhaeuser would supply KiOR with wood chips as the biomass feedstock and KiOR would ship biofuels to Chevron refineries for blending with petroleum-derived fuels. However, negotiations for the deal ended when KiOR declined to supply pilot-plant biomass conversion data or allow Chevron's technical staff to observe and collect pilot-plant samples of crude bio-oil and samples of upgraded gasoline and diesel fuel for further testing. Chevron became concerned with KiOR's ability to deliver bio-oil

BETTING THE FARM

Venture capital financing to fund start-ups and early-stage companies is a risky business for the high failure rate, but it has huge payoff potential if the enterprise turns out to be successful. A decision by an investor to go all in and "bet the farm," or risk losing the investment, can however be built on vetting the technology to reduce the risk. In this case study on KiOR, it is shown that the vetting must be based on truthful information and realistic projections, and that caution is needed when multiple levels of investing are at play, such as when an already invested venture capital firm is seeking funding from additional sources.

meeting its specifications and that KiOR was not acting in good faith. Chevron has since moved on and is now selling a renewable biodiesel product.

KiOR was also interested in obtaining an offtake agreement for its fuels with ExxonMobil. But KiOR data analyzed by ExxonMobil staff left the company doubting that KiOR would be able to deliver and questioning the validity of the data after Loescher's presentation. Earlier, in 2009, Brazil's national oil company, Petrobras, was in discussions with KiOR for possible cooperation in further developing and commercializing KiOR's BCC technology. The company's R&D staff visited KiOR's research facilities in Houston and observed a demonstration of the BCC process at a pilot plant. Subsequently, Petrobras explored biomass to hydrocarbon technology on its own using a similar technology to BCC, through its R&D center known as Cenpes, with disappointing results published later on [123]. In all, each of these companies wanted to work with KiOR, and offered R&D assistance to improve the technology and production facilities. They provided KiOR a golden opportunity to establish itself as a prosperous biofuels company, yet KiOR's managerial approach stood in the way.

During the Q&A session following his presentation in Chicago, Loescher indicated that KiOR was expecting to increase the bio-oil yield to 80 to 90 gallons per dry ton of biomass at a second 500-ton biomass per day plant to be built in Mississippi. This Columbus II plant, which would double KiOR's capacity, would cost $225 million to build and start up, Loescher stated. He noted that the plant would duplicate the "validated" BCC technology used at the existing Columbus I plant.

One of the reasons, it seems, behind Loescher using the inaccurate bio-oil yields in his tcbiomass presentation, was that during this time, July to November 2013, President and CEO Cannon and lead investor Vinod Khosla had engaged additional consultants to evaluate KiOR's performance and its business plan. They needed positive independent evaluation reports to support a prospectus distributed to potential new investors and to attract potential new refinery joint-venture partners and customers for the fuel products.

However, the consulting companies preparing these reports were still using bio-oil yield and production cost projection data provided to them by Cannon and Loescher. To that end, their new reports would simply echo the prior announcements and forecast a promising picture that did not exist for the Columbus facility. But this time the message to investors and the public was more effective, because it was provided via independent technology assessments coming from reliable third-party experts, such as the design engineering firm R. W. Beck, which later became part of SAIC's Energy, Environment, & Infrastructure LLC, now known as Leidos Engineering.

Even though KiOR was backed into a corner by the public announcements of inaccurate plant performance data, the company still raised up to $85 million for the new plant's construction. Most of this money came from Khosla, with $15 million committed by Bill Gates, to be paid in two installments upon reaching certain project milestones, which subsequently were never achieved. Gates noted,

> *"I was impressed when I visited KiOR's Columbus facility and learned more about the company's technology. I am happy to be joining the other investors in support of KiOR's efforts to move its technology forward."*

Following Gates's endorsement, Khosla added,

> *"Khosla Ventures and I have reviewed independent reports on the assessment of the technology and conducted our own significant due diligence as part of this commitment. We are pleased to invest in KiOR with Gates Ventures in this equity financing for the Columbus II project. I believe that KiOR's technology for production of cellulosic biofuels can not only serve as the foundation for a successful and sustainably profitable long-term business but can also scale because of the*

hundreds of saw, pulp, and paper mills that have been shut down and have local feedstock available, providing a much more stable and less price volatile feedstock than oil, while fueling the world's transportation requirements with significantly less geopolitical risk and greenhouse-gas emissions on a life-cycle basis. I expect, as the technology matures over the construction and operation of multiple facilities, it will achieve cost parity with many traditional oil sources such as new deep offshore projects and oil sands, without subsidies."

These statements were issued in an October 21, 2013, press release, "KiOR Completes Equity Investments Anticipated for Columbus II Project," and published in a variety of news outlets. The press release naturally included a forward-looking disclaimer, or so-called standard safe-harbor statement, warning anyone to not get too excited in case the plans did not work out and relieving the company and lead investors of potential liability.

Khosla's comments belie data from KiOR, KBR, and CPERI test results, which Bartek had presented to him in 2009 documenting that, at that point, the BCC technology wasn't working or moreover wasn't going to work in its current form. Khosla was only partly to blame still for this rosy picture: KiOR's executive team had injected the false narrative on bio-oil yields and production costs of finished transportation fuels, despite repeated briefings from the technical team to the contrary, and appears to have muted any dissent within the company, for example when newly hired Director of Technology De Deken raised concerns in summer 2008, when Bartek and Stamires raised concerns in 2009, and when newly hired Chief Operating Officer William Coates raised concerns in summer 2011.

Khosla perhaps was overrepresented on KiOR's Board of Directors, being a member himself along with Samir Kaul, a biochemist with an MBA degree, specifically representing the Khosla Ventures unit known as Khosla Ventures Acquisition Co. II. The unusual manner of Khosla as investor and involvement with more day-to-day management of KiOR raises an ethical concern in hindsight, as it is not clear what Khosla Ventures might have known or not known regarding internal KiOR decisions and how that influenced public fundraising for the new plant.

An informative example is when Khosla directed Andre Ditsch, KiOR's Vice President of Strategy, to change numbers in the company's financial model outside the sphere of the rest of the executive team and the Board of Directors, which came out during a State of Mississippi's lawsuit to recover funds from the bankrupt KiOR [6]:

> "'On January 31, 2011, three weeks after [a] strategy session with Vinod Khosla was conducted,' the state alleges, 'Andre Ditsch notified Yuan Wang, a KiOR employee under Ditsch's direction, that the current yield in the Company's financial model had changed ... Ditsch made this change in the financial model in order that KiOR would appear to potential investors to be commercially viable without RIN and tax credits.' The state added that 'Ditsch did so at Khosla's direction,' the first indication, if true, that Vinod Khosla may himself have become entangled in the faking of yields."

In a *60 Minutes* television episode in January 2014, "The Cleantech Crash," [124], correspondent Lesley Stahl asked a number of questions of Khosla. At the time, Khosla was considered the "father of the cleantech revolution" for the number of nonfossil start-up ventures he created. Even as KiOR was struggling and soon to declare bankruptcy, Khosla described a surprisingly unrealistic picture of KiOR's performance:

> *Khosla: Nature takes a million years to produce our crude oil. KiOR can produce it in seconds.*

> *Narrator: The company took over this old paper mill, where logs are picked up by a giant claw, dropped into a shredder and pulverized into woodchips.*

Khosla: And we take that, add this magic catalyst—

Stahl: This is the secret sauce?

Khosla: Yeah.

Stahl: You throw that on top of the chips?

Khosla: And then, out comes something that looks that looks just like crude oil.

Narrator: The crude is created through a thermochemical reaction in seconds. And by using wood instead of corn, this biofuel doesn't raise food prices which was a concern with ethanol.

Khosla: It smells like crude, it works like crude, except it's 100% renewable. Then it's distilled onsite into...

Stahl: Clean gasoline?

Khosla: Clean green gasoline.

Stahl: This goes right into the tank, right? You don't have to build a new infrastructure

Khosla: Absolutely.

Stahl: You make it sound almost – sorry – too good to be true. There must be a downside.

Khosla: There is no downside.

Khosla's responses to Stahl's questions were understandably simplified for a general audience, but still they did not accurately represent the fundamental problems with KiOR's technology and its poor biofuels results. The reality is that KiOR did not have a "magic catalyst." If the catalyst used at the Columbus plant was indeed magic, then certainly it was the wrong kind of magic for a technoeconomically viable biofuels process.

Stahl went on to explain that there were glitches in KiOR's process, and holes in Khosla's story, that the technology was not perfected and the company was in the red. Khosla downplayed that dilemma, noting that glitches are normal in business start-ups, that nine out of ten go under, and that he expected half his energy companies would fail. Yet, he had just stated there were no downsides to KiOR's process, even though based on prior briefings from KiOR's technical staff he had to know that there were.

Khosla reacted angrily to the *60 Minutes* report, and in an open letter stated that the reporters had cherry-picked facts and fundamentally did not understand how innovation works [125]: "At Khosla Ventures, we invest in companies that have high failure probabilities, but the wins far outweigh the losses." In the letter, Khosla rebutted the reporting on the bioenergy industry generally, but noticeably did not dispute the pending failure of KiOR.

In a *GeekWire* article reporting on the open letter [126], Robert Rapier, a chemical engineer who often writes about the energy industry, was quoted:

> "Vinod Khosla is very smart, but would you let him operate on your heart? No, because it is not his area of expertise. [Vinod Khosla] set up a system where he overpromised and underdelivered, and so the public and the politicians all developed unreasonable expectations."

In a later *Washington Post* article, Rapier echoed those comments and added: "This leads to public perception that advanced biofuels are a boondoggle" [127].

Despite Khosla's role with KiOR, he deserves to be properly

acknowledged for his visionary approach to help clean up our planet and look out for our future. Yet, he was in a precarious situation for having to stay the course with the KiOR technology in place, or follow the proposals for changing to a more realistic approach, initially proposed to him starting in 2009.

In the later part of 2013, a group of KiOR managers and staff became more concerned about the discrepancies in what the management was disclosing to the public relative to the actual results. Some employees who worried about KiOR and their own futures started to leave the company. One example is John Karnes, the Chief Financial Officer since 2011, who had become convinced that the KiOR management's technoeconomic claims were unreasonable. In August 2013, Stamires met with Karnes at his office and showed him plant data with bio-oil yields in the low 20s gallons per dry ton of biomass and estimated production cost close to $6.00 per gallon. Stamires asked Karnes to try to convince Cannon to stop spreading inaccurate information to the public and to form a task force to immediately change the technology. Karnes promised to do his best, and did try, to no avail, before he resigned later in 2013.

During this period of the ongoing exodus of key personnel, Khosla brought O'Connor, the former Chief Technology Officer, who was still on the Board of Directors, back to the CTO role with the specific assignment to fix the technology. O'Connor's recall was not a surprise, because he was considered to be the inventor of the BCC technology and developer of its hydrotalcite (HTC) "magic" catalyst. However, Stamires and others who were convinced the approach would not work, continued working with Cannon on seeking a solution. In October 2013, Stamires provided Cannon with the latest demonstration unit and Columbus I commercial plant results, both indicating low fuel yields and unacceptable high production costs. However, Cannon did not accept the advice to take corrective action. Additionally, Cannon denied that false information was being disclosed by him and his management team to investors and the public.

Stamires reminded Cannon about the conversation they had in

January 2012, when Stamires proposed forming a special task force that he would lead, "Team Oil Yield," which Cannon thought at the time was a good idea and that he would think about it. Cannon informed Stamires that he was still considering forming the task force, but was not sure how to do it without angering Hacskaylo, the Vice-President of R&D.

Subsequently, having failed to convince Cannon to act, and following a letter from Cannon on October 9, 2013, in which Cannon rejected Stamires' evidence that the BCC technology was not working, Stamires decided to take his concerns directly to Vinod Khosla. Unable to set up a meeting, Stamires instead joined O'Connor who was already communicating with Khosla at that time. In May 2014, both visited Khosla at his office in California and presented new proposals describing an operational cost reduction program and technology improvement steps to increase bio-oil yields. Meanwhile, Khosla asked both Stamires and O'Connor to return to KiOR and work together with the technology team on their "Save KiOR" proposals.

When they returned to KiOR in Houston, Cannon and the company's attorney informed Stamires that as a consultant he was not allowed to visit KIOR facilities, only O'Connor was allowed to do so. At face value, this action suggested that Cannon and KiOR's leadership team were opposing Khosla's decision. This action was the beginning of the end for KiOR, effectively the final seed of discord between Cannon, O'Connor, and the KiOR R&D team members, as changes O'Connor proposed to make at the Columbus plant had already been tested before and it was clear they were not going to work. Most members of the Houston-based R&D team declined to cooperate with O'Connor, even for trying to forge a compromise agreement, because, yet again, he was not accepting their pilot plant and demonstration plant test results or listening to their concerns. The KiOR management team supported the R&D team in this regard, as none of O'Connor's proposed changes were put into action, and he resigned from the KiOR Board in August 2014.

But still, Cannon continued to speak publicly using the inflated results and stating that all was going well for the company in meeting its goals, for example in a series of results-earnings calls in 2011-2013, for which transcripts are available [128]. Yet, word was already out that KiOR was struggling [129]. In November 2013, following the Columbus II announcement, Tristan R. Brown wrote about the dangers of relying on next-generation biofuel cost estimates. Brown presented an analysis of the poor financial performance of biofuel companies Amyris and Gevo, which employ engineered microorganisms as catalysts to produce biofuels, and the poor performance of KiOR, with its traditional catalyst approach [130]. Brown, at the State University of New York College of Environmental Science & Forestry, extensively studied technoeconomic performance of biofuel and biochemical production.

The early assessments, Brown indicated, did not always consider the high degree of uncertainty that is inherent in novel pathway technologies:

> *"Production costs presented as point estimates and narrow ranges do not adequately account for this uncertainty and should be given diminished weight by investors performing their due diligence. … No matter how rigorous the methodology underlying the stated estimates, they are in pertain to technologies that have yet to be demonstrated at scale and are therefore very uncertain."*

Indeed, KiOR's seven-year life span, at a cost of close to $1 billion, validates Brown's counsel. Specifically, Columbus I's performance indicated a production cost of several dollars per gallon of crude bio-oil above what was expected, rendering the overall operation overpriced and far from competitive. This is not to say that one day this cost might be bearable, should biomass to biofuels become global society's only option. But at the time the money was being raised and spent at KiOR, the technology problem was ignored, and nothing was done to stop the machine while it was running and going in the wrong direction.

Maxx Chatsko reflected upon this scenario in *Motley Fool* [131]. Chatsko, an experienced scientist with a background in bioprocess engineering, followed the technological and financial developments of KiOR and published several articles. Chatsko mentioned the math, that the 500 ton-per-day Columbus I plant would need to achieve 80 gallons of transportation fuels per dry ton of biomass, or the previously noted 13 million gallons of fuel annually, to meet Khosla Ventures' goal for its return on investment. The plan to add the Columbus II facility would double the numbers, increasing the biomass processing capacity to 1,000 tons per day, or 26 million gallons of fuel annually. Chatsko stated,

> *"It seems counterintuitive to take a struggling platform and make it better by increasing production, but new manufacturing facilities almost always encounter problems. While that may provide a perfectly adequate explanation for a slow start at Columbus and seem to support further expansion, it is just one piece of the puzzle for investing in the company. KiOR is extremely dependent of debt to finance its current operations and its future. The potential may be enticing, but trust me KiOR is too risky for you."*

The truth surfaced a few months after Brown's and Chatsko's assessments, when Cannon, KiOR's President and CEO, revealed for the first time to investors and the public, during a January 9, 2014, Bloomberg business update call, the failure of the Columbus I plant to produce sufficient volumes of biofuels at a competitive cost. Cannon disclosed that Columbus I was producing biofuel with a yield in the low 30s gallons per ton of dry biomass, which was less than half of the 72 gallons claimed just months earlier by Loescher. Even Cannon's number was still higher than the actual 22 gallons reported by plant process engineers.

Internal disruptions at KiOR began to spread more broadly outside the company, to the bioenergy-related business community and to Wall Street. The price of KiOR's shares fell precipitously, and several lawsuits against KiOR's management and Khosla Ventures were filed by investors, the State of Mississippi, and the U.S. Securities & Exchange Commission. Some verdicts have been returned. Cannon, KiOR's President and CEO, as part of a settlement agreement, was ordered to pay a $100,000 fine to the SEC in September 2016. KiOR/ Khosla Ventures paid out $4.5 million for an investor class action lawsuit. The Mississippi lawsuit for the state to recover loans and other costs is still in litigation and it doesn't look like the state will come out breaking even.

5.1 RELYING ON LAB DATA TO GAUGE COMMERCIAL BIOREFINERY PERFORMANCE

Abundant studies continue to be reported in which small-scale bench-top equipment or pilot plants are used to test and evaluate different pathways for converting a variety of biomass and waste hydrocarbon materials to fuels and chemicals. The data often are used to extrapolate and forecast technoeconomic feasibility and even profitability of commercial plant operations. With KiOR's experience behind us, one can't help but wonder how these new studies might differ in their ability to accurately predict future commercial performance.

For that purpose, the results from CPERI using its 1 kg/hour dry biomass capacity pilot plant were especially chosen as an independent source for comparison with KiOR's Columbus I commercial operation. The data set represents a unique situation in which CPERI's pilot plant and KiOR's pilot plants, semiworks unit, and Columbus I commercial plant (Figure 13) all had the same reactor design, and furthermore used the same process, catalysts, and biomass feeds. The CPERI results were later published, offering the opportunity to better compare them with KiOR's actual commercial performance [132].

CPERI's tests were conducted using low-metal beechwood and a commercial grade ZSM-5 catalyst, with the process optimized for maximum bio-oil production. The bio-oil yields were 68 gallons per ton dry biomass containing 27% oxygen and 48 gallons per ton of dry biomass containing 18% oxygen, with a catalyst replacement rate of 2% of the inventory.

These bio-oil yields were included in CPERI's technoeconomic modeling evaluation forecasting the performance of a 500 ton-per day plant, including construction and operation costs, as well as the cost of financing. The expectation from the forecasted performance of a 500 ton-per-day plant producing 48 to 68 gallons of crude bio-oil per ton of dry biomass was for the cost to be in the range of $3.00 to $4.00 per gallon. However, the substantial cost of the hydrocracking and hydrotreating catalysts and the hydrogen needed to upgrade crude bio-oil containing 27% oxygen to transportation fuels was not included.

The construction cost of KiOR's Columbus I plant, including the CFB reactor, woodchip processing yard, shipping facilities, and hydroprocessing system, was close to $230 million. Recall that KiOR eventually shipped about 1 million gallons of biofuel to customers, with Columbus I producing bio-oil in the low-20s gallons per dry ton of biomass with oxygen content close to the needed cutoff of 15%

Figure 13. KiOR's Columbus, Mississippi, 500-ton dry biomass per day commercial plant, with the single-reactor riser tower at center.

when using a commercial ZSM-5 catalyst. The catalyst replacement rate was more than 6%, when 1% is typically acceptable. The catalyst cost was in the range of $8,000 to $9,000 per ton, and remaining oxygen removal from the crude bio-oil required two separate hydrotreating stages to produce a light hydrocarbon product suitable for blending with petroleum-derived fuels. KiOR's actual overall cost to produce gasoline, diesel fuel, and jet fuel at Columbus 1 was more than $6.00 per gallon—the price at the pump for using the fuel on its own or blended with conventional fuels to recover the cost would be higher.

KiOR's $1 billion experiment demonstrated the fundamental inability of the one-reactor, in situ thermocatalyic biomass to bio-oil conversion process with conventional FCC catalysts doubling as heat-transfer media to economically produce an adequate volume of usable bio-oil at commercial scale. We have to keep in mind that, without taking that next risky step as KiOR did, one might never know whether a technology will be commercially successful. This is the nature of advances in science and technology and seeing them through to their fruitful application. We also need to be mindful of the need to have solid reproducible results to decide whether to take any steps—forward, backward, or sideways. The moral of this story is that caution must be taken and the risks considered to ensure the reliability of extrapolating small-scale lab pilot-plant results to predict the technoeconomic performance and sustainability of commercial-scale production of biofuels and biobased chemicals.

5.2 ECONOMIC OUTLOOK

No doubt fuel prices will continue to fluctuate, for many supply, demand, and geopolitical reasons. And this is truly a global issue. Disruptions in supply and demand can arise suddenly from anywhere. It is not out of the realm of possibility that any combination of events could trigger another oil crisis like in the 1970s, or the opposite, a glut of fuels.

Consider that just over a decade ago prices at the average neighborhood gas station in some parts of the U.S. exceeded $5.00 per gallon, reaching a U.S. record average of $4.11 per gallon in 2008, and prices were higher elsewhere in the world, in particular in Europe. Since then, technology has evolved to include greater use of hybrid and fully electric cars, tempering fuel demand. We also know that energy prices came back down, owing to an improved economy and unanticipated record levels of U.S. crude oil and natural gas production, coupled with less U.S reliance on coal; while coal is on its way out globally, supply from China, India, Indonesia, and Australia is still strong, and consumption in China and India is also still strong. Transportation fuel prices for regular gasoline dipped below $2.00 per gallon in some parts of the U.S. in early 2019, before rising again to between $2.20 and $3.65, or a national average of about $2.60, by fall 2019.

After years of success, a pullback on spending in late 2019 by U.S shale fracking producers, who were struggling to make money, hints at a softening of the energy industry. The U.S. stock market recently enjoyed an impressive 11-year bull run through early 2020, but what goes up, must come down. The stock market was perhaps already inflated anyway and needing a correction, but the global COVID-19 coronavirus outbreak in 2020 dealt a heavy blow to the world economy, affecting employment and, critically to this story, putting heavy brakes on the supply and demand for petroleum, the world's busiest commodity market.

Other recent impacts include the on-again, off-again political tensions around the world. Consider Iran, for example. Conflicts include embargos on Iranian oil exports, an Iranian-backed attack on Saudi Arabian oil infrastructure in September 2019, and a tit-for-tat missile attack and counterattack by the U.S. and Iran in January 2020. The political situation in Venezuela has led to similar impacts. Oil exports from Iran and Venezuela, both major producers (Venezuela leads the world in known oil reserves and Iran ranks fourth behind Saudi Arabia and Canada), could be impacted for years. New peace-oriented relationships between countries in the Middle East in late 2020 introduced yet another variable with uncertain impact. The return of the Taliban leadership in Afghanistan in August 2021 could affect that country's oil production, and civil war in Yemen could affect significant undeveloped petroleum resources in that country. Trade between the U.S.—the world's largest economy—and much of the world since 2016 has been tentative. And perhaps one final straw in this haystack of activity is China's growing efforts to build influence in the Middle East for insurance on future petroleum supplies.

In March 2020, as global demand dropped, Russia and Saudi Arabia engaged in a price war that drove the price of oil as low as $31 per barrel, which many economic and business experts would have thought impossible. It does not matter whether oil is that cheap if no one needs it or wants to buy it. But if $31 oil was unexpected, even less expected was the price of oil falling further by April 2020, even dipping to below $0 per barrel as suppliers had nowhere to offload shipments and futures trading required suppliers to temporarily pay refiners and storage facilities to take the oil. And one final variable is continued use of government subsidies to support oil and gas exploration, extraction, and production, as well as government financial support for biobased industries [133-134].

Fuel prices remained low through mid-2021 because of reduced demand, and surprisingly the stock markets recovered in summer 2020 and continued to do well, with the major U.S. indexes hitting new records throughout 2021. As the world slowly began returning to a semblance of normality from the COVID-19 pandemic at the beginning of 2022, even as new variants of the virus were showing up, demand for transportation fuels began increasing, and with it a general inflation in consumer prices and threat of a global recession. U.S. prices ticked up by the end of 2021, to about $80 per barrel of crude oil and to a national average of about $3.30 for regular gasoline. Adding into the mix, the U.S. experienced an episode of short-lived panic buying—another one of those scary fuel availability moments, not unlike in the early 1970s—when a cyberattack on Colonial Pipeline in late April 2021 disrupted fuel distribution along the East Coast.

But this was only a prelude, as fears of a potential Russian invasion of Ukraine in early 2022 further prompted some U.S. refiners to seek substitutes to Russian fuel oil, a common choice of feedstock in gasoline and diesel production. So when an actual invasion came to pass starting in in February 2022, the price of oil jumped to about $115 per barrel, and the U.S. average price for gasoline set a new record, approaching $4.20 per gallon. Major fuel companies severed ties with Russia, pulling out of joint projects in the country, and perhaps most critically shutting down the Nord Stream 1 natural gas pipeline system and mothballing the Nord Stream 2 system that had been waiting to start up.

Natural gas is more important than crude oil for Russia, but oil still plays a major role in the Russian economy. Russia ranks eighth globally in known oil reserves, and supplies about 8% of the world's crude oil, with 60% going via pipeline to Central and Eastern Europe, 20% to China, and 7% to the U.S. Being cut off from exporting to Western countries may not impact Russia much, or for long, because Russia is maintaining carefully crafted relationships with China and India, the world's most populated countries, seemingly planned ahead for continuing trade despite the aggression in Ukraine.

Oil prices continued to climb to about $120 per barrel and to surpass an average $5.00 per gallon of regular gasoline in the U.S. in June 2022. Prices toward the end of 2022 gradually settled back down to what might be a new normal of $70-$80 per barrel, and at the time of this writing in early 2023 a U.S. average of $3.42 per gallon. This lower price was driven by several factors, including a decision by the so-called OPEC-Plus countries, which includes the 13 OPEC members and 10 other oil-producing countries, to cut production to increase demand, the U.S. countering by tapping into its strategic oil reserves, the European Union agreeing to a price cap of $60 per barrel for Russian oil to control prices while ensuring continued reliable supply and limiting Russia's ability to use oil revenue to fund its war, and finally an agreement for Chevron to begin producing oil in Venezuela for the first time since sanctions were levied against that country in 2017. This comes after the Nord Stream pipelines from Russia to Europe in the Baltic Sea started leaking in early October, with sabotage a possibility—it appears explosives were used to damage the pipelines, maybe by a pro-Ukrainian group, but it is uncertain who was responsible. The leaks stabilized, and Gazprom, the Russian company, said delivery could be made through one pipe of Nord Stream 2, if all parties involved could come to an agreement. Still further to complicate the situation, natural gas is five times more expensive in Europe than in the U.S.

While people (in the U.S) now complain about spending $100 to fill up their (extra large) vehicle fuel tanks, we must keep in mind that in 2022 the average American lived in a home that costs $375,000, drove a vehicle that costs $42,000 (and drove about three times as many miles per month than in 1970), and ate out an average five times per week. Then many commiserate because they can't afford to buy groceries and other essentials and care properly for their children and elder family members. Something will have to give, and it might be petroleum-based gasoline and diesel, and this might happen because paying for gas is one of the most visceral ways people feel prices and inflation in their daily lives.

Big Oil's star has already faded significantly, from 2008 when it made up about 15% of the S&P 500, the major U.S. stock market index, to 2020 when it made up about 2.3% of the S&P 500. As oil companies in late 2020 were struggling with underutilized refinery capacity and lower revenues, they lowered their forecasts for future profits and reduced plans for capital expenditures; investment in oil and gas development fell 20% in 2020, according to OPEC. Oil has shrunk as part of all major economies globally. These events are sparking a new line of battle. In May 2021, the International Energy Agency said investment in new fossil-fuel supply projects must immediately stop if the world is to cut net carbon emissions to zero by 2050, a key goal outlined in the 2021 Intergovernmental Panel on Climate Change Assessment Report. In a counterpoint, OPEC announced in October 2021 that the world should be prepared for energy shortages unless there is a global investment in new oil and gas development. Indeed in the U.S., and let us remind that it is still the world's largest petroleum market, oil companies are retiring oil refineries and are not planning on building new ones. Instead, the U.S. federal government is putting significant investment into building electric car battery factories. Some other countries, such as Uzbekistan, are counting on natural gas-to-liquid fuel ventures to supplant petroleum, but the long-term prospects are uncertain.

WHY THE FUSS OVER FUEL COST?

Gasoline and diesel fuel prices in the U.S. depend on several factors: The cost of crude oil; refining costs and profit; distribution and marketing costs and profits, which is different for independent stations and chains such as grocery stores that sell fuel; and state and federal taxes. These factors in turn are under the influence of supply and demand. A certain psychology has evolved about shopping around for the cheapest fuel price, even just to save a few cents per gallon, which almost always seems silly because driving further to save money is likely offset by consuming more fuel. People who tend to worry more about fuel prices do not live in urban areas with good public transportation nor do they own the most fuel-efficient cars, though where one lives and the type of vehicle one drives is a freedom of choice. In other words, fuel prices do not affect everyone equally. Any way one looks at the cost of fuel, the average retail price of a gallon of regular gasoline in the U.S. in 1970 was $0.36, which has the same buying power at the end of 2022 of $2.73, according to the U.S. Bureau of Labor Statistics. Since the early 1970s, gasoline in the U.S. has experienced an average inflation rate of 5.06% per year, compared to the overall inflation rate of 3.94% during the same period. So the gas costing $0.36 in 1970 would cost $4.69 in late 2022. The average cost of regular unleaded gasoline in December 2022 was $3.32 (down from a record high average of $5.01 in June 2022) … less than in 1970 on a constant-dollar basis, but for the reasons noted above (and cheaper than two other commodity fluids—milk and beer).

These actions leave an opening for biofuels and biobased chemicals. It is clear that an economic shift from energy to the tech sector has occurred, and part of that tech sector is new energy technologies. Refiners globally are already studying how to retool for biofuels, such as processing soybean oil, used cooking oil, and animal fat into renewable diesel, to take advantage of tax credits and low-carbon fuel standard credits. For example, Marathon, Phillips 66, Valero, HollyFrontier, and CVR Energy have announced new projects to increase production of renewable fuels. In other developments, BP agreed in October 2022 to spend $4.1 billion to acquire Archaea Energy, a leading U.S. producer of "renewable" natural gas recovered from landfills, a sign that BP continues to move away from traditional sourcing of hydrocarbons for energy production—that is, fossil fuels and traditional drilling methods. Chevron announced just after the Russian invasion of Ukraine that it would buy Renewable Energy Group as part of its plans to invest in green energy. With tax credits, the cost at the pump for this fuel is comparable to that of petroleum-derived diesel. The catalysts and processes developed at KiOR may help. The volume of biodiesel produced in the U.S., about 2.6 billion gallons annually, is still a fraction of the 40 billion gallons of diesel consumed in the U.S. currently. Even with potential for growing market share, the price and availability of sugar, soybeans, and beef tallow will impact future success, as it does for petroleum.

The current situation likely does not signal the end of the age of petroleum. Consider that petroleum resources remain abundant, and as technology advances, extraction methods will improve further, for example, potentially making drilling for oil in the Arctic Ocean practical one day. But as bioenergy production grows and fuel consumption has temporarily waned, we are at an inflection point on the future pathway for biobased fuels and chemicals to compete against fossil fuels. The conclusion to draw at this juncture in reading this account is that, yes, one can produce fuels and chemicals from biomass, but not yet in the large volumes needed at competitive production costs, and with low- or net-zero carbon emissions. Taking all into consideration, these technoeconomic challenges have become much greater and much more complex on the road toward achieving a globally sustainable bioeconomy. However, this situation may change soon, out of necessity, and another lesson we are learning is that the technology will need to be much better to ensure success of an economically viable commodity biofuels business when it is needed.

One reason to continue to anticipate this need comes from recognizing that global population continues to increase and will approach 10 billion people by 2050, assuming recent trends in lower reproduction rates do not alter this projection. And as the world is becoming further developed and modernized, we are consuming more energy per capita. Even with technology advances that will enable more efficient use of fuels and electricity, energy demand is only going to increase in the coming decades. Short of complete reliance on solar energy, which might include controlled nuclear fusion technology if it can be advanced to a usable scale, we need to be vigilant and act with a bit more vigor to ensure we will always have ample amounts of clean, affordable liquid fuels, which will gradually require using every available resource, including virgin unprocessed and processed biomass, waste biomass, and materials such as waste plastics, waste paper and cardboard, solid municipal waste, and recycled glass.

Whether or not any doomsday scenarios play out, these conditions suggest the urgent need to finish what we have started and develop new processes with decarbonization capabilities to effectively convert most of the globally available, nonfood, low-cost biomass sources to commodity clean fuels and/or chemicals, even as we still have enough petroleum-derived fuels and chemicals. Integrating biobased fuel and specialty chemical production with electricity production, similar to how Brazil has done, while capturing greenhouse gas emissions, is a promising approach. As KiOR's lesson shows us, these processes must be scalable to large commercial plants while being cost-effective, environmentally acceptable, amenable to "global unwarming," economically sustainable, and ideally without government subsidies. It is a tall order.

5.3 LESSONS LEARNED FROM KIOR'S EXPERIENCE

To that end, it is worth contemplating the valuable experiences we have learned from KiOR's series of unfortunate events and outcomes, and put it all to good use. Overall, KiOR's documented history provides an account of the spectrum of critical technical and managerial steps a company should avoid.

The foremost action in KiOR's timeline, which got the company off on the wrong foot, was ignoring early red-flag warnings on failing pilot-plant results that demonstrated the BCC technology was underperforming and unscalable to commercial plants. In 2008, early in KiOR's trajectory, Director of Technology De Deken, following his technology review, pointed out to the management team that rushing toward demonstrating the BCC technology at a multi-barrel-per-day scale without reproducible corroborating experimental data, while publicly trying to create corporate value, is a recipe for failure. Despite his advice, the management, led largely by O'Connor, because Cannon had been hired by O'Connor and was still new on the job and not up to speed on all aspects of KiOR's technology, disregarded De Deken's recommendation to change the technology. De Deken's actions could be used as a textbook on the ethical basics of starting, growing, and managing a business. The reality was that De Deken was forced to leave the company after only five months, and his potentially valuable contributions to KiOR's success were lost.

De Deken's red flag was confirmed in late 2008 by Bartek, the Senior Manager of Process & Catalyst Development, who further validated the limitations of the technology and echoed De Deken in telling KiOR's leadership the BCC technology was failing and the company must do something radically different to save the project. The whipping-the-dead-horse mentality of continuing to pursue the failing approach rather than replace it, even with subsequent warnings from Stamires and technical staff members, gradually whittled away funding to a point of no return.

A second action that spurred on KiOR's demise was the management's message of inflated biofuel yields and deflated production costs to cast the company in a favorable light to attract new investors. However, the day of reckoning came on January 9, 2014, when Cannon was forced to announce that KiOR's Columbus I plant could not come close to producing the volume of biofuels anticipated and committed to be delivered to customers. KiOR filed for bankruptcy later that year.

Another of KiOR management's disruptive actions involved Coates, the Chief Operating Officer, who had been hired in June 2011. Coates was an experienced oil industry executive who soon discovered that the management team was overstating company performance. For stepping forward to rectify the situation by reporting to the Board of Directors how the leadership team had "cooked the books," Coates was dismissed after only five weeks on the job, and with a nondisclosure agreement as part of his severance package, he remained publicly silent, allowing the company to continue operating on its charted failing course. KiOR's stock price continued rising, reaching a peak capitalization of $2 billion in October 2011.

Still another of KiOR management's mistakes involved senior

chemical engineer Max Ross, which is a bit more complex and requires some explanation, as it gets to the heart of the matter of KiOR's pathway to failure. Ross had 30-plus years of experience operating, modifying, and revamping Exxon's and later ExxonMobil's oil refinery FCC units worldwide in its petroleum and heavy-oil refinery business, with some of the unit designs similar to the Columbus I plant. His technical qualifications and operational refinery facility experience could not possibly have been a better match to KiOR's needs. Ross joined KiOR in June 2013 and worked in the Houston office, but he spent most of his time at the Columbus plant working with the FCC-type reactor trying to increase the bio-oil yield with an acceptable oxygen content—his further work might have turned things around for KiOR.

In July 2013, shift senior chemical engineer Charlie Zhang working at the Columbus plant gave Ross a copy of the technical proposal prepared by the stealth team that had been presented earlier to KiOR's leadership. Ross met with Stamires at the Houston office in July and August, and they discussed the production data Ross had obtained while he was working at the Columbus plant and compared it with the plant data Zhang was providing to Stamires. They agreed that the hydrotreater was working well and quality biofuels were being produced. The oxygen content was acceptable in the range of 10 to 15%; however, the bio-oil yields were only in the low 20 gallons per dry ton of biomass and the catalyst replacement rate was too high.

They also agreed that a computer process simulation study from 2011-2012 by Agnes Dydak, a chemical engineer with oil refinery process operational experience, had correctly predicted the bio-oil yield at the Columbus plant to be in the low 20s—again, one-fourth the target amount. However, KiOR failed to benefit from Dydak's fortune-telling, which went hand-in-hand with Ross's and Zhang's findings, because Loescher in his supervisory role did not allow the results to be shared within the company, and further ordered the records to be destroyed, so that the results would not contradict the management's inaccurate bio-oil yield claims.

Nevertheless, by August 2013, KiOR had no money or time to change the reactor setup and biomass devolatilization process to that described in the stealth team's proposal. Instead, as a quick and temporary fix, a two-particle catalyst blend system as described earlier was considered, but it was never tested at Columbus I. The blend consisted of a low-activity, high-heat-capacity nonzeolitic material developed at KiOR, plus a smaller portion of a catalyst containing a high-SAR phosphated ZSM-5. Both Ross and Stamires and separately Zhang submitted new engineering and process plans to Loescher to improve the plant's bio-oil yield while reducing costs. But Loescher did not act on these proposals.

On August 29, 2013, Stamires went to Cannon and presented him with the latest Ross and Zhang plant production data, which overall demonstrated an economically unfeasible operation. Stamires told Cannon that he believed it was his duty to inform the Board of Directors about the Columbus plant's unacceptable results, and he would ask for the Board's assistance to change the technology. In response, Cannon thanked Stamires and told him that he would think about how to do it with the blessing of Hacskaylo, the Vice President of Research & Development.

Stamires never heard back from Cannon. On September 25, Artzer, the Vice President and General Counsel, notified Stamires that his consulting contract, expiring the following month, would not be renewed due to operational cost-reduction measures. In reality, the decision to not renew the consulting contract stemmed more from Stamires's persistence in trying to change the technology and management team, which will be explained later on. KiOR's management offered Stamires a substantial nondisclosure financial payout for agreeing to keep all of KiOR's activities and information confidential, and to not take his case to the Board of Directors. Stamires declined the offer.

Meanwhile, Stamires and other KiOR executive staff members who were objecting to the recent decisions had learned from outside sources, to their surprise, that KiOR's management had secured funding and planned to proceed with the project to double the Columbus I plant processing capacity, to a total of 1,000 tons per day. The new Columbus II plant was estimated to cost about $225 million, although that figure later grew to $400 million. This was good news for some: investors and the stock market reacted positively. The company's leadership continued with its bullish plant production results and announced that it would be easy enough to duplicate the "already commercially proven technology."

Several news articles reported the KiOR development [135]. Cannon stated,

> *"The Columbus II project marks an important step in the execution of the long-term business plan of KiOR."*

Cannon added that,

> *"... we believe that this project will enable us to achieve profitability in 2015 ... that construction timing and cost are more certain for the Columbus II project, as it is essentially a duplicate of our existing Columbus facility that can be leveraged to reduce construction risk ... we plan to achieve operational and technological synergies between the two Columbus facilities, as we expect to incorporate our most recent technology developments into both the new Columbus II facility and retroactively to the existing Columbus facility ... and we expect a shorter startup period for the Columbus II facility as a result of sharing personnel, infrastructure, and operational knowledge with the existing Columbus I facility. This expansion of Columbus has been partially enabled by significant improvements to our technology ... ."*

Stamires, concerned with this decision to cast another quarter-plus billion dollars into a failing technology, decided to become more involved in the Columbus II project. On October 1, 2013, as his contract was expiring, he sent a letter to Cannon and the management team offering to work with Ross and Zhang for the next six to eight months without compensation, with the first step to change the biomass devolatilization and catalyst technologies. However, Cannon and the management team declined his offer.

Cannon's decision points to an additional critical set of managerial actions taken by KiOR's leadership team: A concerted effort to keep the Board of Directors in the dark on the company's progress and performance and to prevent the Board from taking corrective action. In large part, the Board relied on fellow member O'Connor, being an inventor of the BCC process, as a trusted source of information.

In particular, in early 2012 the Board of Directors informed Cannon that it wanted to send O'Connor to KiOR's corporate office in Houston on a fact-finding mission and to have a comprehensive technology review meeting. The Board asked Cannon to arrange the meeting, which he did for March 7-8. However, KiOR's management team of Cannon, Hacskaylo, Loescher, and Artzer resisted this audit-like intervention by the Board. These executives appear to have aligned to make sure O'Connor would not learn anything beyond the information they had been officially providing on KiOR's business model and technology performance, including actual biofuel production yields, costs, and catalyst test results. This team preselected a few topics and technology areas to be discussed with O'Connor, as well as hand-selected whom O'Connor would be allowed to see. The presentations were scripted to prevent disclosing contradictory information—in other words, O'Connor's meeting was staged to make it a fact-hiding mission.

Recall that Stamires had previously arranged for Vasalos to share data, sign an agreement with KiOR to license his pilot-plant design, prepare research reports, and visit KiOR on several occasions to consult on the technology developments. Ahead of O'Connor's visit to Houston, Stamires tried to convince Cannon to allow Vasalos to attend. Stamires informed Cannon that he had already discussed the nature and the purpose of the meeting with Vasalos, and that Vasalos was willing to participate. Stamires reiterated to Cannon that it was critical to have an independent expert familiar with the technology—and already bound by a confidentiality agreement—at the meeting to make the findings and derived conclusions credible to the eyes of the KiOR Board members, and thereby investors and potential customers, and help inform them on corrective actions to take. O'Connor was also trying to convince Cannon to invite Vasalos. Yet, on February 22, Stamires and O'Connor were informed that Cannon had denied their submitted proposals for including Vasalos in the forthcoming audit and technology review on the basis of legal confidentiality concerns.

At the end of the March 2012 meeting, it became clear that the management's actions were effective in preventing O'Connor from getting the information he was seeking on behalf of the Board of Directors, and that the management was in effect violating its fiduciary duties to the Board. On March 13, Stamires met with Cannon and expressed his disappointment with how O'Connor's visit had turned out, in particular for dodging O'Connor's questions. However, Stamires managed to make sure that O'Connor did not leave Houston and return to the Board of Directors with his pockets empty.

Stamires met privately with O'Connor twice during his stay in Houston to transfer key information, with a mutual agreement that O'Connor would deliver it to the Board of Directors and assist the members in understanding and appropriately using it. This transferred package included R&D progress reports describing new catalyst preparations and pilot-plant testing; new and more efficient pilot-plant reactor designs that increased the bio-oil yields; Loezos' pilot-plant test results and analysis; Vasalos' analysis and evaluation of CPERI's pilot-plant test results using KiOR's catalysts, biomass feed, and sand testing for comparison with published competitor results; Vasalos and McGovern's independent expert report; and Stamires' stealth team proposal describing a new technology for producing higher biofuel yields at lower costs. However, O'Connor failed to deliver this mission-critical information to the Board members.

Although the Board was largely in the dark, it had gotten wind of the management missteps from members Samir Kaul of Khosla Ventures and Gary Whitlock, Chief Financial Officer of CenterPoint Energy. Stamires had informed intellectual property attorney Jennifer Camacho about KiOR's management problems, and she did in turn inform Kaul. In addition, before Coates was dismissed, he had informed Whitlock about the management problems. The idea was also floated that Whitlock, as head of the Board's Audit Committee, should interview Stamires to find out what was going on at KiOR and report back. However, Whitlock never contacted Stamires.

Stamires at this point took on a whistleblower role and brought his concerns directly to KiOR's Board of Directors by contacting Board Member William J. F. Roach, a Ph.D. metallurgist with oil and gas industry experience [6]. Roach arranged a meeting attended by attorneys from WilmerHale LLP, representing the KiOR management team, and attorneys from Locke Lord LLP, representing the Board of Directors. The meeting took place in Houston on Friday, January 17, 2014, just a few days after Cannon first publicly disclosed some of KiOR's troubles.

Stamires presented a comprehensive account of his findings regarding the Columbus I plant's performance, including the management's inflated bio-oil yields and deflated production costs—the proverbial "cooking the books," which by this time had turned into "overcooking the books"—indicating his belief that their statements were part of an intentional plan. In particular, Stamires emphasized the need to immediately replace the management team to enable a fundamental revamp of the technology so that the overall process could become technoeconomically feasible and competitive.

Following the Houston meeting, two attorneys from Locke Lord visited Stamires near his home in Newport Beach, California, where Stamires had his complete set of KiOR work documents with detailed information. They spent two and one-half days discussing further KiOR's fundamental problems and another half day copying documents for the Locke Lord attorneys to prepare a report for the Board of Directors to use in deciding how to address the Columbus plant technology and changing the management. Stamires had passed many of these documents to O'Connor in their private meetings in Houston on March 7-8, 2012, with the agreement that O'Connor would share them with the Board, though in the two years leading up to January 2014 O'Connor did not share the information with the Board as agreed. Following the meeting in California, Stamires and the attorneys had several follow-up conversations regarding the information and discussed these new details with some Board members. However, by the time the Locke Lord attorneys had finished preparing their KiOR fact-finding and assessment report with recommendations and submitted it to the Board, KiOR had run out of money and stopped operating.

About that time, in January 2014, Ross was dismissed from KiOR for a safety violation, accused of entering an area of the plant without having a permit, which seemed innocuous, but nonetheless made any further efforts to change the technology to save KiOR essentially impossible. In the end, KiOR's Board was impeded twice by its own members, when O'Connor failed to deliver the documents Stamires had given him and when Whitlock did not follow up on Coates' request to contact and discuss KiOR's technology problems with Stamires.

One final set of actions that warrants summarizing is O'Connor's contributions to KiOR's demise, as he had a hand in most of the company's activities during its existence. In 2007-2008, while O'Connor was Chief Technology Officer, he overruled the R&D staff's recommendations to calibrate the pilot plants using sand as a heat-transfer medium, to compare pilot-plant performance with the published data of competitors, and to test and evaluate the performance of new and lower cost catalyst compositions outside the existing BCC Technology. On another front, at the end of 2008, before he left the CTO position, O'Connor prepared KiOR's 2009 R&D plan in which he left out any projects for calibrations and for testing promising new biomass conversion technologies and high-performance, low-cost catalysts.

O'Connor had lost creditability with members of the R&D staff, as he effectively put them in a hot seat for losing their jobs when he was CTO. O'Connor made things worse when KiOR staff discovered that, after he left the CTO position but was still a member of the Board of Directors, he had filed for a patent as sole inventor on Dec. 28, 2011, claiming new materials, catalysts, and biomass conversion processes [136]. Some of these items actually had been invented earlier by KiOR R&D team members and formally documented in the company's R&D records and intellectual property portfolio. This new patent cut KiOR out of the picture altogether by assigning the technology to BIOeCON, where O'Connor was still Managing Director and major shareholder.

O'Connor carried out this "technology swipe" without notifying KiOR's management, the R&D staff, or the company's intellectual property management team to investors and the public. O'Connor directed KiOR's IP team while he was the company's CTO, giving him control over what was patented. That was at a time when O'Connor was declining the technical team's requests to calibrate reactors and to test new catalysts, perhaps as a cost-saving measure as described earlier, even though as CTO he was in charge of financially supporting BCC work by KiOR contractors, raising an ethical concern about his motives for personal gain. And after he left the CTO role but was still on the Board of Directors and in frequent contact with the R&D team, he remained aware of all of KiOR's new patentable developments. Furthermore, the patenting activity occurred when confidentiality and company non-competitive agreements existed between O'Connor, BIOeCON, and KiOR. In essence, O'Connor's patenting activity cost KiOR a valuable IP asset at a crucial time when the company was trying to resolve its technology issues and stay afloat, plus it set up a potential patent interference in which O'Connor or BIOeCON could legally challenge KiOR's technology as it evolved. The key point in this discussion is that the patented technology is out there, much of it currently unused.

KiOR's key technical personnel became upset when they learned about the patent from outside sources, adding to the frustrations

HOW PATENTING WORKS

A patent is an exclusive right granted for a product or a process that provides a new way of doing something or offers a new technical solution to a problem. To get a patent, technical information about the invention must be disclosed in a patent application. Once a patent is granted, generally by a national government or regional agency, the owner may give a license to others to use the invention on mutually agreed terms or sell the rights to someone else who becomes the patent owner. Once a patent expires, usually after 20 years, the protection ends and the invention enters the public domain, where anyone can use the product or process; the patent owner could possibly extend patent coverage by getting additional patents on improvements. Patents are territorial, so that rights are only applicable in the country or region in which a patent has been filed and granted. Patent rights are usually enforced in a court, which has the authority to stop infringement, such as someone else trying to patent the same idea or to use the invention without permission, with the patent owner having the main responsibility for monitoring the patent and taking action against infringers. For example, a patent interference arises when pending patent applications, or a pending patent application and an unexpired patent, contain overlapping claims. A patent derivation court proceeding determines whether someone derived the invention from someone else and/or the earlier application was filed without authorization.

they already had with O'Connor and the company's leadership. In hindsight, O'Connor, who is Dutch, suggested that cultural differences between Dutch and American stakeholders, in particular with the Houston operations, prevented an open corporate dialogue and might have led to misunderstandings [6]. Although this "culture clash" appears to have been more perceived than real, it has contributed to confusion and misunderstanding of the publicly documented details of what really led to the failure of KiOR's technology and the company's demise, overshadowing the lack of accountability. This confusion and misunderstanding in essence impeded development of a sustainable and environmentally acceptable biomass-to-biofuels technology, a disservice to scientific advancement.

O'Connor appeared to remain optimistic about his BCC technology and the future of biofuels, despite KiOR's legacy. In an interview published January 24, 2017, with Tammy Klein, a lawyer and journalist who is a biofuels industry consultant, he lamented that his original idea for producing bio-oil at BIOeCON was set aside by the KiOR leadership team in Houston [137]. In this interview, he repeats some comments made to other reporters after he was forced to resign from KiOR's Board of Directors in 2014. O'Connor states that the KiOR establishment assumed that it had more knowledge about the BIOeCON conversion strategy than the people like himself who had developed it.

> *"It is very strange. They thought they had invented their own process and so at that moment KiOR took a kind of a different turn. I think it's important for me to say this because on the technology side I believe it still can work. But they took a short cut and they neglected what many people now see as the most important part which is the biomass pretreatment step. … They kept on the same road, and yes, it unfortunately hit the wall. So that's very sad. Not only for the people there, but for all of us, for the whole industry because I think—and that's the thing which hurts me the most—that it has hurt the whole industry. People say, oh you know, you see, it can't be done, and that is the wrong conclusion. I think that's the wrong conclusion."*

This series of events in KiOR's story is illuminating for how companies operate in general, that is in disclosing how information, truthful or not, is conveyed internally, or not. People do make mistakes and sometimes fail to see something obvious happening. There is a psychology to some of our errors, especially regarding anticipation and expectations. But preventable or correctable mistakes cease to be errors when they are intentionally overlooked. A key value of experience is that we grow to be wary of such potential flaws. It is possible that KiOR's leadership knew something regarding the technology's potential for success that they never shared with the technical team, but the results at Columbus I stemming from rigorous preliminary development indicate otherwise.

In late 2014, KiOR's Columbus I plant stopped operating and the company declared bankruptcy. KiOR did survive for some time through several reincarnations. The company changed its name to Mard Inc. (an interesting choice, perhaps meaning to caretake or standing for Mean Absolute Relative Difference), and later became Inaeris Technologies, a private company financed by Khosla Ventures that remained under the same management seeking investors. Later losing most of its employees, and Khosla funding, Inaeris started selling its assets: the Columbus plant in Mississippi and all research equipment located in Houston. In May 2019, Inaeris posted notice of an auction selling remaining equipment and company assets, and a once-promising company was finally defunct. Yet, KiOR maintains a pulse, still recognized as Mard on paper as a business entity in Delaware, last noted in a bankruptcy court filing in October 2020 [138]; BIOeCON also appears to have gone out of business, or at least into hibernation.

5.4 POSTMORTEM ANALYSIS

Starting in 2015, KiOR postmortem analyses began to appear, typically focusing on studying entrepreneurship and technoeconomic performance. These reports were sometimes inaccurate or misleading about what actually went wrong at KiOR. For example, one case study published December 7, 2015, by the Digital Initiative: Technology & Operations Management, an online forum produced by Harvard Business School, states [139],

> *"However, two subsequent decisions did not allow the business and operational model to coexist well. Firstly, as the company tried to scale up, they hired dozens of researchers with Ph.D.s, while not recruiting people with operational backgrounds who actually had facilities experience. KiOR was in essence, a network of very smart chemistry researchers. This was an operational decision that severely impacted the company's ability to execute its business vision."*

Contrary to the above statement, KiOR's business model was based on projections made with misleading yield and production cost data. In addition, many of the key technical personnel who operated and managed KiOR's biomass to bio-oil operations and its upgrading to fuels from the beginning in 2007 through 2014 were all extensively qualified chemical process engineers. Most of them had long, hands-on oil-refining operational experience working in major petroleum refining companies. Only two had doctoral degrees, De Deken and Loezos, and they came to KiOR from major oil companies. This technical staff early on in 2008-2009 and through to late 2013 had shown that the BCC technology had inherent fundamental technical limitations—the approach was not capable of being scaled up to produce an acceptable volume and quality of bio-oil at a competitive cost necessary to sustain a commercial business.

Harvard's assessment, diagnosis, and conclusion of KiOR's problems erred in not mentioning the consistently failing technology or the company's failure to broker deals with Catchlight (Weyerhaeuser-Chevron), ExxonMobil, and Petrobras stemming from distrust of KiOR's management. This analysis actually followed an earlier Harvard Business School case study on KiOR [140]. The earlier analysis erred in not accurately portraying how KiOR was formed, the early failing lab results, and the lack of proof-of-concept of the BCC technology. While the Harvard studies can be discarded by those in the know, as reports by a normally trustworthy institution they nevertheless have an unfortunate detrimental effect by misleading the public, scientists and engineers, investors, and the industry on what actually went wrong at KiOR, impeding further advancement of science and technology.

Another KiOR postmortem example sowing blame, "Three Critical Lessons We Can Learn from Failed Startups," appeared in 2016 in the Huffington Post's "What's Working: Small Business" [141]. The article states:

> "O'Connor said that the absence of people with real technical experience running energy facilities 'hurt KiOR a lot,' impeding operations and mismanaging resources."

A similar quote appeared in articles from other news outlets. This statement is notable because it points to O'Connor as the possible source of information in the Harvard case studies, but also goes against O'Connor's comments elsewhere [6] in which he mentions that the technical team had made excellent progress in building the organization, scaling up the BCC process and proving the BCC concept, and developing improved catalysts under his leadership when he was KiOR's Chief Technology Officer from 2007 through 2009.

> "During this period, the team progressed from the pilot to the demo phase and then commercial phase at the Columbus plant in a record time of less than five years, considered impossible in the process industry."

O'Connor had managed the KiOR R&D work and oversaw the

hiring of technical personnel, some of whom were his prior colleagues and friends. O'Connor, who owned a significant number of shares of KiOR, remained on the Board of Directors with a strong influence over company operations, even after leaving his role as CTO, before he resigned from the Board in May 2012 and again when he rejoined the Board in early 2014 before involuntarily resigning on August 31, 2014.

In O'Connor's 2016 comments, that the absence of people with real technical experience running energy facilities hurt KiOR, he was suggesting the company was imbalanced, with too many catalyst people and not enough refinery process people. Additionally, with regard to his comments on proving the concept, two research labs in Europe, a Petrobras pilot plant in Brazil, and two pilot plants in the U.S. all consistently validated that the BCC technology was not working—they actually reproducibly invalidated the BCC concept as it was formulated and put into use. Regarding leading the research into improved catalysts, O'Connor as CTO had consistently impeded the KiOR research team from pursuing new R&D work to replace the failing BCC technology, including calibrating the pilot plants and testing new catalysts.

O'Connor was one of the most, if not the most, knowledgeable person on all of KiOR's technology, and its technology problems, by the time the Harvard Business School report came out in 2015. Yet, he does not appear to have challenged its accuracy and/or correct the report, and seemingly he used it as a personalized alibi to deflect attention from his crucial role in KiOR's failure.

This inaction follows the earlier situation when O'Connor did not appear to try to correct information passed on to Gates and Khosla when they visited the Columbus plant and were making investment decisions. Even though O'Connor noted in a 2012 technology assessment that the yields from the KiOR process had "not improved considerably over the past two years," and in his August 2014 Board of Directors resignation letter wrote that he had been "shocked" by the lack of progress and that the company had disregarded his recommendation to take a "drastically different approach," he had resisted changes requested from other members of the technical team in favor of his original BCC ideas. Whereas O'Connor's impact significantly contributed to leaving the company in critical condition, in his resignation letter O'Connor pointed elsewhere for the company's failings [142]. This blame was vigorously disputed by KiOR's management as being "false, misleading, or inappropriate," and further accused O'Connor of withholding information, for unauthorized communications disclosing confidential information, and possibly violating the company's insider trading policy.

In retrospect, this postmortem account of the life and times of KiOR is overdue. The benefit of this story is hopefully to provide insight to those who may have been treated unfairly when trying in earnest to save the company, to the rest of the employees who endured the difficult circumstances and were required to recalibrate their careers when the company went bankrupt, and to investors who lost their money and are entitled to know all the causes behind KiOR's failure.

All told, the most critical goal for presenting this story is to benefit those working in new start-up companies and future energy development efforts. We hope this account aids advancement of the underlying science and technology, encourages next steps by the investment community, and paves the way for public acceptance to drive those advances in developing a globally sustainable, environmentally acceptable, and commercially successful biomass and waste conversion to biofuels and biochemicals industry.

5.5 COMPARISON OF BIOMASS-TO-BIOFUELS COMPANIES

In the spirit of postmortem analysis, this is a good time to take a moment to compare the technologies and operations of some other biomass-to-biofuel companies. One with a similar experience as KiOR is Range Fuels. This company had a goal of using forest industry waste to generate synthesis gas (called "syngas," a mixture of

H_2, CO, and other gases such as CH_4 and CO_2 used as a feedstock) and further produce C_1 to C_6 mixed alcohol fuels in a nonfermentation process. Like KiOR, the company, which operated from 2007 to 2011, ultimately failed and closed down. One hindsight analysis suggests that Range Fuels succumbed because of a combination of factors, including reliance on a flawed technoeconomic analysis that underestimated feedstock cost, overestimating catalyst performance, miscalculating the challenges of sizing up the technology to commercial scale, and an imbalance of too many managerial level staff and not enough production staff [143]. In some ways, a scenario similar to KiOR's.

Yet not all companies share this experience. Take Ensyn, for example, which originated in the mid-1980s and continues operating its Rapid Thermal Processing (RTP) biomass conversion process. This is an approach similar in some technological respects to KiOR's failed approach—Ensyn and KiOR were major competitors. In 2012, Ensyn formed a joint venture with oil-refinery technology firm Honeywell UOP, named Envergent Technologies, with the purpose of pooling their respective experiences and intellectual property to produce clean transportation fuels from waste biomass. Ensyn has a record of success in producing bio-oil from cellulosic biomass. UOP is the world leader in developing and licensing oil-refining catalysts and process technologies—in essence, an ideal partner to provide the know-how to convert crude, oxygenated bio-oil produced using the Ensyn technology to clean, fully deoxygenated light hydrocarbon transportation fuels.

So far, Envergent seems to be a successful joint venture supplying bio-oil fuel for power plants and for industrial process heating, as well as using one of the side products as liquid smoke, a food flavoring and condiment. Envergent has also been successful in recruiting customers for its new generation of gasoline, diesel fuel, and jet fuel produced from bio-oil via UOP's hydroprocessing technology. The company has several commercial operations in place, including a facility in Quebec, Canada, that according to company press releases is projected to convert 65,000 metric tons of dry biomass forest residue per year to

10.5 million gallons of bio-oil when fully operational. These numbers were announced in 2016, but no updated information appears to be available, nor a date for full onstream production. However, an Ensyn patent indicates the process has a lower yield than that mentioned in the press release, leaving open speculation as to what may really come to pass, reminiscent of KiOR's story [144-145].

Envergent has yet to demonstrate economic sustainability and profitability at commercial scale without subsidies, which will remain challenging because of the need to use low-quality, low-cost biomass rather than high-quality, low-metals biomass and to minimize the cost of supplying hydrogen and hydrotreating catalysts for upgrading the crude bio-oil. Another company that led the way in the new wave of biomass-to-fuels development in the 1990s based on the use of CFB reactor systems was Dynamotive Energy Systems, but the company was unable to sustain its efforts in developing a commercial production facility and has gone out of business. One observation is that joint ventures would seem to be more viable for successful biofuels production over standalone companies.

Further regarding Range Fuels, it sold its facility to LanzaTech, which is working toward commercial success in converting biomass-derived syngas into alcohols by a fermentation process [146]. Gevo appears to be making progress in its engineered microorganism approach to make biofuels, but the company continues to operate at a loss. Some market analysts predicted the Gevo might reach a breakeven point in 2022, but in its most recent earnings report the losses were continuing to grow. Enerkem has partnered with Shell to convert municipal waste into syngas then use Fischer-Tropsch chemistry to produce fuel-type compounds. Other companies have changed focus. For example, Amyris is now centered on specialty and performance chemicals, flavors and fragrances, cosmetics ingredients, pharmaceuticals, and nutraceuticals.

In another example, DuPont spent $200 million developing a cellulose-based fermentation process at a plant in Iowa to convert corn cobs, stalks, and leaves into ethanol, but the plant closed in 2017 because the economics were not working out. Verbio bought the facility from DuPont and is converting it into a traditional corn-based ethanol facility and expanding it by adding fermenters to make natural gas from the corn plant residuals, with the leftover material to be returned to corn fields as a soil amendment. On another front, Phillips 66 is working on a new facility to convert animal fats and cooking oil into diesel and gasoline.

Alder Fuels is a recent entry into the field, started in October 2021 to produce sustainable aviation fuels (SAFs). The company uses microbes to convert food waste to volatile fatty acids and then with zirconium oxide upgrade the fatty acids to ketones followed by hydrotreating with a platinum-on-nickel catalyst to remove oxygen and create jet fuel. A number of other start-up companies are angling to enter the SAF market, especially as regulatory agencies globally seem ready to require fuel suppliers to blend SAFs into conventional fuels, like they do already with ethanol into gasoline [147].

And as a final note, in late 2022 and into 2023 the high price for natural gas has made accessing the fuel profitable by nearly any means, with most landfills investing in infrastructure or negotiating deals with energy companies. For example, Archaea Energy and Waste Management Inc., just a couple of names, have big plans to siphon off the gas and effectively turn what was once just thought of as "the local dump" into a renewable-energy facility.

After KiOR failed, O'Connor continued to be involved in several biobased technologies. For example, O'Connor was later Managing Director at Antecy, a company supported by Bill Gates with a goal to take CO_2 from industrial and power plant flue gases and utilize it to make fuels and chemicals and in other applications; in September 2019, Antecy was bought and merged with Swiss firm Climeworks. Most recently, O'Connor has had a hand in formation of Yerrawa, a company developing technology for catalytic conversion of methane and waste organic materials such as plastics to hydrogen and solid carbon. Yerrawa has filed patent applications, and the company website mentions licensing some technology from other companies, all with O'Connor listed as an inventor. These include Climeworks and

Cellicon, as well as recent patents by O'Connor and others that are assigned to BIOeCON, which still exists as a holding company. The Yerrawa website also lists among the company's intellectual property two patents assigned to KiOR in which O'Connor and Stamires are among the inventors, raising questions about who now owns the patent rights and who can use the technology.

Time will tell, but KiOR's approach to remove most of the oxygen from the crude bio-oil at the front of the process during biomass devolatilization with low-cost catalysts and fewer overall steps, rather than all at the end of the bio-oil upgrading as Ensyn does, seems like a more economically favorable approach. On the basis of KiOR's experience, and more recent developments, we recommend the use of low-cost catalysts discussed in this book, especially those prepared from natural minerals, to prepare assorted catalyst particles exhibiting basic, acidic, or dual-functionality for next-generation technologies. Beyond biomass, these catalysts should be further useful in cost-effective conversion of soybean or other plant or animal oil and fat to fuels and chemicals.

Resources are plentiful for future efforts to expand biofuel production and usage, beyond crude oil, sugars, and biobased oils, should the prices and availability of those resources become unfavorable in the future. Carbon monoxide, carbon dioxide, and methane, obtained from renewable resources and waste, especially from power plants, petrochemical plants, industrial manufacturing plants, and landfills, can be used as a feedstock or for generating electricity or for residential use. For example, new developments in electrochemical techniques for selective and energy-efficient production of light hydrocarbons from CO or CO_2 using solar, wind, hydroelectric, or nuclear power are waiting to happen, should challenges in construction, grid infrastructure, demand, and high-capacity battery storage be overcome. In fact, companies such as Dow are on a quest to use small, modular nuclear reactors to generate electricity, rather than using fossil fuels, to power chemical plants such as carbon-neutral ethylene crackers. Optimized, low-cost catalysts for mediating this chemistry will be needed.

6.0 SUMMARY

We have been on a roller coaster ride since those days of the 1970s oil crisis when long lines at the pump were driving us crazy, and the ride is not over—as 2022 ended, we still felt angst over fuel prices and availability. From a historical perspective, KiOR's costly failure is significant, as the funding from investors and taxpayers could have been better spent. One could argue that the $1 billion gambit was necessary, because without such an expensive experiment one could never have known the truth about the technology's promise, nor gained the valuable lessons learned.

In a December 15, 2015, article in *Fortune,* journalist Katie Fehrenbacher, who often wrote about KiOR, summed it all up regarding the company's fate [148]:

> *"But if the size of the failure equals the magnitude of the educational payoff, then KiOR will have been a very valuable lesson indeed."*

The catalyst and process technology topics discussed here will hopefully be helpful to those working in the field and contribute to the advancement of biofuels and biobased chemicals development and sustainable protection of our planet (Figure 14) [149-151]. Unused patents from BIOeCON, KiOR, and others exist and are waiting to be used for low-cost production of net-zero carbon energy. We need to keep in mind also that new knowledge from better understanding of biomass to clean fuels conversion continues to accumulate [152-153].

Figure 14. Chevron is now selling a new biofuel in California, "Renewable Biodiesel B20," which is a blend of about 80% renewable biodiesel made from hydrotreated vegetable oil and/or animal fat and 20% biodiesel (fatty acid methyl ester) made from transesterified vegetable oil or animal fat. This development has historical context for KiOR, which had a promising agreement a decade ago with Catchlight Energy, a Weyerhaeuser-Chevron joint venture, to sell biofuels to Chevron refineries, but never came to fruition (photo by Jackson Donahoo, grandson of coauthor Dennis Stamires).

In particular, a commercially attractive means of converting lignin into fuels and chemicals remains a key challenge. Lignin is the most recalcitrant part of biomass, comprising as much as 30% dry weight of the material. Finding an optimized approach to utilizing it along with the cellulose and hemicellulose, other than burning it, holds enormous promise on the road to a sustainable future [154-157]. Another key need is to find an optimized approach for converting recyclable plastics such as polyethylene and other solid municipal waste carbon materials into light hydrocarbon fuels or chemicals [158-159]. Carbon resources are additionally plentiful as an extension of fossil fuel and biomass use, especially emissions from power plants, petrochemical plants, and industrial manufacturing plants.

Right now, petroleum and fossil fuels are available, and although subject to price variability, are still relatively cheap. However, this situation will not be the case forever. Big Oil still has a lot of sway: Although most companies have publicly pledged to transition to greener energy, beneath the surface they still are solidly focused on petroleum and working toward a future for natural gas over a biomass economy, as disclosed by a U.S. House of Representatives Committee on Oversight and Reform investigation in 2021-2022. Yet at some point, biomass conversion to generate green energy and carbon-neutral or carbon-negative biofuels to meet society's needs can reasonably be expected to be more attractive, both financially and to better protect the environment. The protections include continuing to curb ozone depletion and reducing the rate of global warming and its effects such as the loss of Earth's glaciers—in fact to undo it, that is, to help the world achieve global unwarming—and thereby limiting the effects of destructive climate change including loss of plant and animal species and thus biodiversity.

This environmental protection is not just from reducing greenhouse-gas emissions stemming from extracting, refining, transporting, storing, and burning petroleum, coal, and natural gas, which includes leaking abandoned wells, but also from greenhouse gases escaping from landfills. In turn, environmental decarbonization and protection can be financially rewarding, by finding ways to capture, store, and utilize CO, CO_2, CH_4, and other gases, as well as the problematic nitrogen and sulfur compounds associated with them [160], as feedstocks for value-added commercial products.

One additional environment-related fuel issue is the ongoing use of lead compounds as anti-knock agents and octane enhancers in fuels. While lead has largely been eliminated from automobile fuels for decades, it is still included in fuel for piston-engine aircraft. In the U.S. it amounts to more than 180 million gallons consumed per year. Automobile fuel once was a leading source of lead pollution in the atmosphere. According to the U.S. Environmental Protection Agency, aviation fuel is now the primary source. EPA is moving toward requiring lead-free aviation fuel, a regulatory decision among other things designed to protect the health of children who live near airports. This move will require new fuel formulations, which could rely on biofuels.

We do have to keep in mind that these processes are energy intensive and have to be designed well. And the time to cash in could be soon: Reducing regulatory barriers and stimulating economic growth following the COVID-19 outbreak could be a new type of catalyst. Indeed, in 2021-22 the U.S. government has been mulling over details in massive spending bills to improve transportation infrastructure coupled with addressing climate change. And 2022 was a milestone for decarbonizing the world's energy systems as the first year in which investment in solar and wind projects, electric vehicles, and other technologies is thought to have equaled global investment in fossil fuels, which is about $1.1 trillion [161].

New interest in expanding a global bioeconomy as an insurance policy for our supply of liquid transport fuels and key chemicals had already been under way [162]. In one example, new bipartisan legislation is under consideration in the U.S. Congress with "The Bioeconomy Research & Development Act of 2020" [163]. And even further, California in September 2020 announced a ban on the sale of new fully gasoline- and diesel-powered cars by 2035, a decision supported further by a U.S. federal proposal in August 2020 aiming for half of new car sales to be zero-emission vehicles, presumably targeting electric vehicles but some net-zero biofuel vehicles, by 2030. In California, the new rule allows for battery-powered or hydrogen fuel cell vehicles or hybrid-electric vehicles with batteries that run for at least 80 km; currently about 15% of new cars sold in California are electric. The California announcement is significant because the state is the largest auto market in the U.S., and more than a dozen other states typically follow California's lead when setting their own auto emissions standards. Electric cars are coming—most if not all carmakers are committing to producing fewer internal combustion engine vehicles and more electric vehicles to meet growing demand—and we can reasonably expect all new passenger vehicles will be electric one day, though transportation fuels will still likely be needed for trains, airplanes, ships, construction equipment, and long-haul trucking.

In addition, at the end of 2020, U.S. Congress passed the Sustainable Chemistry Research & Development Act as part of the National Defense Authorization Act for Fiscal Year 2021 to support manufacturing and jobs while also protecting human health and the environment by helping to realize the full innovation and market potential of sustainable green chemistry technologies [164]. Still further, in September 2022 U.S. President Joseph Biden issued an Executive Order "Advancing Biotechnology and Biomanufacturing Innovation for a Sustainable, Safe, and Secure American Bioeconomy" to encourage federal agencies to work toward a more sustainable future [165].

Perhaps most striking and impactful is the Inflation Reduction Act of 2022, approved by the U.S. Congress in August 2022 [166]. This legislation will deliver the largest investment in climate action in U.S. history, with $370 billion planned over the next decade to rapidly scale up renewable clean-energy production and drive substantial reductions in greenhouse-gas emissions while achieving a net carbon zero or better footprint of transportation fuels. Still it won't be easy: A number of companies have made sustainability pledges and are finding them hard to keep; for example, in 2020 Delta Air Lines vowed to invest $1 billion over 10 years to reduce its carbon footprint with new planes and cleaner jet fuel. Although the airline says it is carbon neutral, it comes with a bit of "greenwashing," because to meet its targets the company is purchasing millions of dollars of carbon offsets so far, and not directly cutting the emissions, thereby limiting the impact. One related development that came in December 2022 was the announcement that researchers at Lawrence Livermore National Laboratory had for the first time successfully achieved a net energy gain from a nuclear fusion experiment, that is more energy was produced than it took to initiate the reaction. There is a caveat with this experiment, that much more energy was needed to power the lasers to get to the initiation point than was produced. This fusion experiment is stage one of a long, many-stage process, not unlike that undertaken by KiOR and other energy companies, to see if fusion energy will ever be practical on a commercial scale—that is, can it be operated in a technoeconomical way that is safe and does not negatively impact the environment.

Bill Gates, who invested in KiOR and continues to invest in many global projects, published a book in February 2021 titled "How to Avoid a Climate Disaster: The Solutions We Have and the Breakthroughs We Need." In reflecting on this work, Gates said [167],

> "I wrote 'How to Avoid a Climate Disaster' because I think we're at a crucial moment. I've seen exciting progress in the more than 15 years that I've been learning about energy and climate change. The cost of renewable energy from the sun and wind has dropped dramatically. There's more public support for taking big steps to avoid a climate disaster than ever before. And governments and companies around the world are setting ambitious goals for reducing emissions. What we need now is a plan that turns all this momentum into practical steps to achieve our big goals."

One can argue that we have the breakthroughs we need in hand. When the time is right, and with the right approach, then, yes, indeed, as described here, the pot of gold at the end of the rainbow is there for achieving the goal of technoeconomically feasible, environmentally acceptable, and sustainable commercial-scale biomass-to-biofuels and -biobased chemicals production.

REFERENCES

[1] Tullo A. The Future of Oil is in Chemicals, Not Fuels. Chem Eng News 2019;97(8):26-29. https://cen.acs.org/business/petrochemicals/future-oil-chemicals-fuels/97/i8.

[2] Finlay M. Old Efforts at New Uses: A Brief History of Chemurgy and the American Search for Biobased Materials. J Ind Ecol 2003;7(3-4):33-46. https://doi.org/10.1162/108819803323059389.

[3] U.S. Biomass Research & Development Act of 2000, https://www.energy.gov/sites/prod/files/2014/04/f14/biomass_rd_act_2000.pdf [accessed March 1, 2023].

[4] U.S. Food, Conservation & Energy Act of 2008, https://www.govinfo.gov/content/pkg/PLAW-110publ234/pdf/PLAW-110publ234.pdf [accessed March 1, 2023].

[5] U.S. Agriculture Improvement Act of 2018, https://www.govinfo.gov/content/pkg/BILLS-115hr2enr/pdf/BILLS-115hr2enr.pdf [accessed March 1, 2023].

[6] Lane J. KiOR: The Inside True Story of a Company Gone Wrong. Biofuels Digest 2016, https://www.biofuelsdigest.com/bdigest/2016/11/24/kior-the-story-of-a-company-gone-wrong-part-5-the-collapse/[accessed March 1, 2023].

[7] Mohorčich J. What Can Biofuel Commercialization Teach Us about Scale, Failure, and Success in Biotechnology? Sentience Institute 2019, https://www.sentienceinstitute.org/biofuels [accessed March 1, 2023].

[8]   Total Energy 2020. U.S. Energy Information Administration, https:// www.eia.gov/totalenergy/[accessed March 1, 2023].

[9]   World Energy Balances 2020. International Energy Agency, https:// www.iea.org/subscribe-to-data-services/world-energy-balances-and -statistics, [accessed March 1, 2023].

[10]   Lane J. Seeing the World from a Different View: The Digest's 2019 Multi-Slide Guide to ABLC's Secret Speaker. Biofuels Digest 2019, https://www.biofuelsdigest.com/bdigest/2019/08/29/seeing-the-worl d-from-a-different-view-the-digests-2019-multi-slide-guide-to -ablcs-secret-speaker/[accessed March 1, 2023].

[11]   Verleger Jr PK. $200 Crude, the Economic Crisis of 2020, and Policies to Prevent Catastrophe. Petroleum Economics Monthly 2018, https:// www.pkverlegerllc.com/assets/documents/180704200CrudePaper.pdf [accessed March 1, 2023].

[12]   Jacoby, M. The Shipping Industry Looks for Green Fuels. Chem Eng News 2022;100(8):22-26. https://cen.acs.org/environment/ greenhouse-gases/shipping-industry-looks-green-fuels/100/i8.

[13]   Frankiewicz T. Process for Converting Oxygenated Hydrocarbons into Hydrocarbons. US Patent 4,308,411, Dec. 29, 1981.

[14]   Choi C K. Pyrolysis of Carbon-Containing Material. US Patent 4,064,018, Dec. 29, 1977.

[15]   Garrett D, Mallan G. Pyrolysis Process for Solid Wastes. US Patent 4,153,514, May 8, 1979.

[16]   Williams PT, Horne PA. Characterization of Oils from the Fluidized Bed Pyrolysis of Biomass Zeolite Catalyst Upgrading. Biomass Bioenergy, 1994;7(1-6):223-236. https://doi.org/10.1016/0961-9534(94)00064-Z

[17]   Scott DS, Piskorz J. (1984). The Waterloo Fast Pyrolysis Process for the Production of Liquids from Biomass. Canada Bioenergy R&D Seminar 1984;5:407-12.

[18]   Scott DS, Piskorz J, Radlein D. Liquid Products from the Continuous Flash Pyrolysis of Biomass. Ind Eng Chem Process Des Dev 1985;24(3), 581-588. https://doi.org/10.1021/i200030a011

[19]   Chantal P, Kaliaguine S, Grandmaison J L, Mahay A. Production of Hydrocarbons from Aspen Poplar Pyrolytic Oils over H-ZSM5. Appl Catal 1984;10(3):317-332. https://doi.org/10.1016/0166-9834(84)80127-X

[20]   Vasalos I. Fluid Bed Retorting and Its System. US Patent 4,404,083, Sept. 13, 1983; US Patent 4,511,434, April 16, 1985.

[21]   Samolada MC, Stoicos T, Vasalos I. An Investigation of the Factors Controlling the Pyrolysis Product Yield of Greek Wood Biomass in a Fluidized Bed. J Anal Appl Pyrolysis 1990;18(2):127-141. https://doi.org/10.1016/0165-2370(90)80003-7

[22]   Lappas AA, Samolada MC, Iatridis DK, Voutetakis SS, Vasalos IA. Biomass Pyrolysis in a Circulating Fluid Bed Reactor for the Production of Fuels and Chemicals. Fuel 2002;81(16):2087-2095. https://doi.org/10.1016/S0016-2361(02)00195-3

[23]   Chambers RW. Apparatus and Method for Recovering Useful Hydrocarbon Products. US Patent 4,235,676, Nov. 25, 1980.

[24]   Ritter S. Tire Inferno. Chem. Eng. News 2013;91(43):10-15. https://pubs.acs.org/doi/abs/10.1021/cen-09143-cover

[25]   Piskorz J, Majerski P, Radlein D. Energy Efficient Liquefaction of Biomaterials by Thermolysis. US Patent 5,853,548, Dec. 29, 1998.

[26]   Biagini E, Barontini F, Tognotti L. Devolatilization of Biomass Fuels and Biomass Components Studied by TG/FTIR Technique. Ind Eng Chem Res 2006;45(13):4486-4493. https://pubs.acs.org/doi/10.1021/ie0514049

[27]   Biagini E, Tognotti L. A Generalized Procedure for the Devolatilization of Biomass Fuels Based on the Chemical Components. Energy Fuels 2014;28(1):614-623. https://pubs.acs.org/doi/10.1021/ef402139v

[28]   Jenson JB. Shaloilogy and Oil-Shale Nomenclature. Chem Metall Eng 1922;26:509.

[29]   ACS National Historic Chemical Landmark 1996. The Houdry Process for the Catalytic Conversion of Crude Petroleum to High-Octane Gasoline, https://www.acs.org/content/acs/en/education/whatischemistry/landmarks/houdry.html[accessed March 1, 2023].

[30]   Stamires D, O'Connor P, Jones W. Mg-Al Anionic Clay Having 3R2 Layer Stacking. US Patent 6,468, 488, Oct. 22, 2002. Newman SP, Jones W, O'Connor P, Stamires DN. Synthesis of the 3R2 Polytype of a Hydrotalcite-like Mineral. J. Mater. Chem. 2002;12:153-155

[31]   Kelkar S, Saffron CM, Andreassi K, Li Z, Murkute A, Miller DJ, Pinnavaia TJ, Kriegel RJ. A Survey of Catalysts for Aromatics from Fast Pyrolysis of Biomass. Appl Catal B 2015;174-175:85-95. https://doi.org/10.1016/j.apcatb.2015.02.020

[32]   Wan S, Waters C, Adam S, Gumidyala A, Rolf J, Lobban L, Resasco D, Mallinson R, Crossley S. Decoupling HZSM-5 Catalyst Activity from Deactivation During Upgrading of Pyrolysis Oil Vapors. ChemSusChem 2015;8(3):552-559. https://doi.org/10.1002/cssc.201402861

[33]   Singer LS, Stamires DN. Trace Ferromagnetism in Zeolites. J Chem Phys 1965;42(9):3299-3301. https://doi.org/10.1063/1.1696413

[34]   Freeman Jr D, Stamires D. Compaction of Zeolites. US Patent 3,213,164, Oct. 19, 1965. J. Subcasky WJ, Place TM, Stamires DN, Anderson WG, Parker Jones HA, Segovia G, Salisbury RG. Development of a High Temperature Battery. NASA Contract NA53 6002, Oct. 15, 1965 https://ntrs.nasa.gov/api/citations/19650027180/downloads/19650027180.pdf [accessed March 1, 2023].

[35]   Breck DW. Crystalline Zeolite Y. US Patent 3,130,007, April 21, 1964.

[36]   Rabo JA, Pickert PE, Stamires DN, Boyle JE. Molecular Sieve Catalysts in Hydrocarbon Reactions, Second International Congress on Catalysis. Platinum Met Rev 1960;4(4):141-143.

[37]   Milton RM. Molecular-Sieve Adsorbents. US Patent 2,882,244, April 14, 1959.

[38]   Huber GW, Jae J, Vispute T, Carlson T, Tompsett G, Cheng Y. Catalytic Pyrolysis of Solid Biomass and Related Biofuels, Aromatics and Olefin Compounds. US Patent 20090227823, Sept. 10, 2009.

[39]   Massachusetts District Court 2010. Case 3:09-cv-30225-MAP, https://www.plainsite.org/dockets/rttk41uh/massachusetts-district-court/kior-inc-v-huber/ [accessed March 1, 2023].

[40] Scott, A. Investors Chase Bioaromatics. Chem Eng News 2020;98(8): 18-19.

[41] Bartek R, Brady M, Stamires D. Controlled Activity Pyrolysis Catalysts. US Patent Application 20120142520, June 7, 2012; WO Patent Application 2010124069, Oct. 28, 2010.

[42] Stamires D. Catalyst Compositions for Use in a Two-Stage Reactor Assembly Unit for the Thermolysis and Catalytic Conversion of Biomass. US Patent 20130261355, Oct. 3, 2013.

[43] Bartek R, Brady M, Stamires D. Refractory Mixed-Metal Oxides and Spinel Compositions for Thermocatalytic Conversion of Biomass. US Patent 8,921,628, Dec. 30, 2014.

[44] Stahl L. The Unlikely, Eccentric Inventor Turning Inedible Plant Life Into Fuel. 60 Minutes, Jan. 6, 2019, https://www.cbsnews.com/news/marshall-medoff-the-unlikely-eccentric-inventor-turning-inedible-plant-life-into-fuel-60-minutes [accessed March 1, 2023].

[45] O'Connor P. Polymeric Material of Photosynthetic Origin Comprising Particulate Inorganic Material. US Patent 8,648,138, Feb 11, 2014.

[46] Brady M, Cordle RL, Loezos P, Stamires D. Two-Stage Reactor and Process for Conversion of Solid Biomass Material. US Patent Application US 20120117860, May 17, 2012

[47] Stamires D, Jones W, O'Connor P. Doped Anionic Clays. International Patent Application WO 2002064504, Aug. 22, 2002.

[48] Stefanidis SD, Karakoulia SA, Kalogiannis KG, Iliopoulou EF, Delimitis A, Yiannoulakis H, Zampetakis T, Lappas AA, Triantafyllidis KS. Natural Magnesium Oxide (MgO) Catalysts: A Cost-Effective Sustainable Alternative to Acid Zeolites for the In Situ Upgrading of Biomass Fast Pyrolysis Oil. Appl Catal B: Environmental, 2016;196:155-173

[49] Sanchez-Valente J, Lopez-Salinas E, Sanchez-Cantu M. Process For Preparing Multimetallic Anionic Clays and Products Thereof. US Patent US20080274034, Nov. 6, 2008.

[50] Stamires D, Jones W, O'Connor P. Doped Anionic Clays. US Patent 7,022,304, April 4, 2006.

[51] Stamires D, Jones W, Daamen S. Process For The Preparation Of Anionic Clay. US Patent 6,593,265, July 15, 2003.

[52] Stamires D, Jones W, O'Connor P. In-Situ formed Anionic Clay-Containing Bodies. US Patent 7,008,896, March 7, 2006.

[53] Zhang X, Sun L, Chen L, Xie X, Zhao B, Si H, Meng G. Comparison of Catalytic Upgrading of Biomass Fast Pyrolysis Vapors over CaO and Fe(III)/CaO Catalysts. J Anal Appl Pyrolysis, 2014;108:35-40. https://doi.org/10.1016/j.jaap.2014.05.020

[54] Stamires D. Caution on Red Mud Use. Chem Eng News 2017;95(21):4. https://cen.acs.org/articles/95/i21/Caution-red-mud-use.html

[55] Ritter S. A More Natural Approach to Catalysts. Chem Eng News 2017;95(8):26-32. https://cen.acs.org/articles/95/i8/natural-approach-catalysts.html

[56] Agblevor F, Halouani K. Catalytic Pyrolysis of Olive Mill Waste. US Patent 20150065762, March 5, 2015.

[57] Agblevor F, Elliott D, Santosa D, Olarte M, Burton S, Swita M, Beis S, Christian K. Brandon S. Red Mud Catalytic Pyrolysis of Pinyon Juniper and Single-Stage Hydrotreatment of Oils. Energy Fuels 2016;30(10):7947-7958. https://doi.org/10.1021/acs.energyfuels.6b00925

[58] Dight LB, Lawrence B, Garcia-Martinez, JG, Valla I, Johnson MM. Compositions And Methods For Improving The Hydrothermal Stability Of Mesostructured Zeolites By Rare Earth ion Exchange. US Patent 8,524,625, Sept. 3, 2013.

[59] Lim J, Brady M, Novak K. Substrates with Calibrated Pore Size and Catalyst Employing Them. US Patent 4,356,113, Oct. 26, 1982.

[60] Stamires D, Brady M. Catalyst Composition With Increased Active-Site Accessibility For The Catalytic Thermoconversion Of Biomass To Liquid Fuels And Chemicals And For Upgrading Bio-Oils. US Patent 10286391, May 14, 2019; WO 2013123297, Aug. 22, 2013.

[61] Wang S, Dou T, Li Y, Dou Z, Zhang Y, Li X, Yan Z. Hydrothermally Stable Aluminosilicate Mesostructures Prepared from Zeolite ZSM-5. J Mater Sci 2007;42:401-405. https://doi.org/10.1007/s10853-006-0767-3

[62] Ludvig, MM. Preparation of MFI-type Crystalline Aluminosilicate Zeolites. US Patent 6,667,023, Dec. 23, 2003.

[63] Lappas AA, Dimitropoulos VS, Antonakou EV, Voutetakis SS, Vasalos IA. Design, Construction, and Operation of a Transported Fluid Bed Process Development Unit for Biomass Fast Pyrolysis: Effect of Pyrolysis Temperature. Ind Eng Chem Res 2008;47(3):742-747. https://pubs.acs.org/doi/10.1021/ie060990i

[64] Lappas AA, Kalogiannis KG, Iliopoulou EF, Triantafyllidis KS, Stefanidis SD. Catalytic Pyrolysis of Biomass for Transportation Fuels. WIREs Energy Environ 2012;1(3):285-297. https://onlinelibrary.wiley.com/doi/abs/10.1002/wene.16L. Lappas AA, Kalogiannis KG, Iliopoulou EF, Triantafyllidis KS, Stefanidis SD. Catalytic Pyrolysis of Biomass for Transportation Fuels. In: Lund PD, Byrne J, Berndes J, Vasalos IA, editors, Advances in Bioenergy: The Sustainability Challenge. Wiley; 2016, p. 45-56.

[65] Thiele WW. Catalytic Conversion. US Patent 2,400,176, May 14, 1946.

[66] Engelhard Corp. Catalyst Compositions. US Patent 5,071,539, Dec. 10, 1991.

[67] Xu M, Macaoay J. Manufacture of ZSM-5-type Zeolites and their Use as Cracking Catalysts. US Patent 6,908,603, June 21, 2005.

[68] Toombs AJ, Armstrong WE. Alumina Adsorbents. US Patent 3,726,811, April 10, 1973.

[69] Brady M, Bartek R, Stamires D, O'Connor P. Comminution and Densification of Biomass Particles. US Patent 8,465,627, June 18, 2013.

[70] Turkevich J, Ikawa T, Nozaki F, Stamires D. Catalytic Activity and Nuclear Radiation. Proceedings of Industrial Uses of Large Radiation Sources, Salzburg, Austria, 1963:41-56. https://inis.iaea.org/collection/NCLCollectionStore/_Public/44/065/44065979.pdf?r=1

[71] Lai R, Yan Y, Gavalas GR. Growth of ZSM-5 Films on Alumina and Other Surfaces. Microporous and Mesoporous Mater 2000;37(1-2):9-19. https://doi.org/10.1016/S1387-1811(99)00188-2

[72] Keliona P, Likun W, Zhang H, Vitidsant T, Reubroycharoen P, Xiao R. Influence of Inorganic Matter in Biomass on the Catalytic Production of Aromatics and Olefins in a Fluidized-Bed Reactor, Energy Fuels 2017;31(6):6120-6131. https://doi.org/10.1021/acs.energyfuels.7b00339

[73] Fahmi R, Bridgwater A, Darvell L, Jones J, Yates N, Thain S, Donnison I. The Effect of Alkaline Metals on Combustion and Pyrolysis of *Lolium* and *Festuca* Grasses, Switchgrass and Willow. Fuel 2007;86(10):1560-1569. http://doi.org/10.1016/j.fuel.2006.11.030

[74] Fahmi R, Bridgwater A, Donnison I, Yates N, Jones J. The Effect of Lignin and Inorganic Species in Biomass on Pyrolysis Oil Yields, Quality and Stability. Fuel 2008;87(7):1230-1240. https://doi.org/10.1016/j.fuel.2007.07.026

[75] Shafizadeh F. Pyrolysis and Combustion of Cellulosic Materials. Adv Carbohydr Chem 1968;23:419-474. https://doi.org/10.1016/S0096-5332(08)60173-3

[76] Shafizadeh F, Furneaux RH, Cochran TG, Scholl JP, Sakai Y. Production of Levoglucosan and Glucose from Pyrolysis of Cellulosic Materials. J Appl Polym Sci 1979;23(12):3525-3539. https://doi.org/10.1002/app.1979.070231209

[77] Gray M, Corcoran W, Gavalas G. Pyrolysis of a Wood-Derived Material. Effects of Moisture and Ash Content. Ind Eng Chem Process Des Dev 1985;24(3):646-651. https://doi.org/10.1021/i200030a020

[78] Raveendrana K, Ganesh A, Khilar K. Influence of Mineral Matter on Biomass Pyrolysis Characteristics. Fuel 1995;74(12):1812-1822. https://doi.org/10.1016/0016-2361(95)80013-8

[79] Oudenhoven S, Westerhof R, Aldenkamp N, Brilman D, Kersten S. Demineralization of Wood Using Wood-Derived Acid: Toward a selective Pyrolysis Process for Fuel and Chemicals Production. J Anal Appl Pyrolysis 2013;103:112-118. https://doi.org/10.1016/j.jaap.2012.10.002

[80] Stamires D, Brady M, O'Connor P, Rasser J. Process for Producing High Quality Bio-Oil In High Yield. US Patent 20120144730, June 14, 2012.

[81] Brady M, O'Connor P, Stamires D. Biomass Pretreatment Process. US Patent 8,168,840, May 1, 2012.

[82] Stamires D, O'Connor P. Pretreatment of Biomass with Carbonaceous Material. US Patent 8,552,233, Oct. 8, 2013.

[83] Mullen C, Boateng A, Goldberg N. Production of Deoxygenated Biomass Fast Pyrolysis Oils via Product Gas Recycling. Energy Fuels 2013;27(7):3867-3874. https://doi.org/10.1021/ef400739u

[84] Bartek R, Yanik S. Coprocessing of Biomass and Synthetic Polymer-Based Materials in a Pyrolysis Conversion Process. US Patent 9,040,761, May 26, 2015.

[85] van der Stelt M, Gerhauser H, Kiel H, Ptasinski K. Biomass Upgrading by Torrefaction for the Production of Biofuels: A Review. Biomass Bioenergy 2011;35(9):3748-3762. https://doi.org/10.1016/j.biombioe.2011.06.023

[86] Bartek R, Brady M, Stamires D. Catalytic Hydropyrolysis of Organophilic Biomass. US Patent 8,063,258, Nov., 22, 2011.

[87] Nolte M, Shanks BA. Perspective on Catalytic Strategies for Deoxygenation in Biomass Pyrolysis. Energy Technol 2017;5(1):7-18. https://doi.org/10.1002/ente.201600096

[88] Iliopoulou E, Antonakou E, Karakoulia S, Vasalos I, Triantafyllidis K. Catalytic Conversion of Biomass Pyrolysis Products by Mesoporous Materials: Effect of Steam Stability and Acidity of Al-MCM-41 Catalyst. Chem Eng J 2007;134(1-3):51-57. https://doi.org/10.1016/j.cej.2007.03.066

[89] Szymanski H, Stamires D, Lynch G. Infrared Spectra of Water Sorbed on Synthetic Zeolites., J Opt Soc Am 1960;50(12):1323-1328. https://doi.org/10.1364/JOSA.50.001323

[90] McDaniel C, Maher, P. New Ultra-Stable Form of Faujasite. Molecular Sieve Conference, Society of Chemical Industry, London, April 1967.

[91] Hansford R. Stabilized Zeolites and their use for Hydrocarbon Conversions. US Patent 3,354,077, Nov. 21, 1967. Kerr GT. Hydrothermally Stable Catalysts of High Activity and Methods for their Preparation. US Patent 3,493,519, Feb. 3, 1970.

[92] Alafandi H, Stamires D. Exchanged Faujasite Zeolites. US Patent 4,058,484, Nov. 15, 1977; US Patent 4,215,016, July 29, 1980. Hydrothermally Stable Catalysts Containing Ammonium Faujasite Zeolites.US Patent 4,085,069, April 18, 1978.

[93] Stamires D. Properties of the Zeolite, Faujasite, Substitutional Series: A Review with New Data. Clays and Clay Miner 1973;21:379-389. http://dx.doi.org/10.1346/CCMN.1973.0210514

[94] Mullen C, Boateng A, Mihalcik D, Goldberg N. Catalytic Fast Pyrolysis of While Oak Wood in a Bubbling Fluidized Bed. Energy Fuels 2011;25(11):5444-5451. https://doi.org/10.1021/ef201286z

[95] Stamires D, Brady M. Mesoporous Zeolite-Containing Catalysts for the Thermoconvesion Of Biomass and for Upgrading Bio-Oils. US Patent Application 20160017238, Jan. 21, 2016; WO 2013123299, Aug. 22, 2013.

[96] Rabo JA, Pickert PE, Boyle JE. Zeolite Molecular-Sieve Catalysts for Hydrocarbon Conversion. US Patent 3,236,761, Feb. 22, 1966.

[97] Stamires D, Brady M. Catalyst Compositions Comprising in situ Grown Zeolites on Clay Matrixes Exhibiting Hierarchical Pore Structures. US Patent Application 20150027871, Jan, 29, 2015.

[98] Brady M, Stamires D. Catalyst System Having Meso and Macro Hierarchical Pore Structure. US Patent 10,286,391, May 14, 2019.

[99] Plank CJ, Rosinski EJ. Selective Cracking Catalyst. US Patent 3,277,018, Oct. 4, 1966. Crystalline Zeolite Composite Catalyst for Hydrocarbon Conversion. US Patent 3,140,253, July 7, 1964. Cracking Hydrocarbons with a Crystalline Zeolite. US Patent 3,140,251, July 7, 1964. Cracking with a Synthetic Zeolite Catalyst. US Patent 3,140,249, July 7, 1964.

[100] Stamires D, O'Connor P, Laheij EJ, Sonnemans M, Wilhelmuset J. Continuous Process and Apparatus for the Efficient Conversion of Inorganic Solid Particles. US Patent 6,903,040, June 7, 2005.

[101] Maher PK, McDaniel CV. Ion Exchange of Crystalline Zeolites. US Patent 3,402,996, Sept. 24, 1968.

[102] Ward WJ. Catalytic Cracking Process Using Ammonia-Stable Zeolite Catalyst. US Patent 4,036,739, July 19, 1977.

[103] Scherzer J. Acid Dealuminated Y-Zeolite And Cracking Process Employing It. US Patent 4,477,336, Oct. 16, 1984.

[104] Haden Jr W, Dzierzanowski F. Microspherical Crystalline Zeolitic Cracking Catalyst. US Patent 3,433,587, March 18, 1969.

[105] Freeman Jr D, Stamires D. Electrical Conductivity of Synthetic Crystalline Zeolites. J Chem Phys 1961;35:799. https://doi.org/10.1063/1.1701219

[106] Rahimi N, Karimzadeh R. Catalytic Cracking of Hydrocarbons Over Modified ZSM-5 Zeolites to Produce Light Olefins: A Review. Appl Catal A 2011;398(1-2):1-17. https://doi.org/10.1016/j.apcata.2011.03.009

[107] Adkins B, Stamires D, Bartek R, Brady M, Hacskaylo J. Process for Converting Biomass to a Fuel. US Patent 9,044,741, June 2, 2015.

[108] Adkins B, Stamires D, Bartek R, Brady M, Hacskaylo J. Catalyst for Thermocatalytic Conversion of Biomass to Liquid Fuels and Chemicals. US Patent 9,649,624, May 16, 2017.

[109] Juttu G, Shogren K, Adkins B. Phosphorus Promotion of Zeolite-Containing Catalysts. US Patent 9,522,392, Dec. 20, 2016.

[110] Pine L. Comatrixed Zeolite and Phosphorus-Alumina. US Patent 4,584,091, April 22, 1986.

[111] Walker D, Schaffer A. Cracking Catalyst and Process using Aluminum Phosphate-Containing Matrix. US Patent 4,873,211, Oct. 10, 1989.

[112] Macedo J. Kaolin Containing Fluid Cracking Catalyst. US Patent 5,082,815, Jan. 21, 1992.

[113] Demmel E. Method for Producing Attrition-Resistant Catalyst Binders. US Patent 5,190,902, March 2, 1993.

[114] Bartek R, Brady M, Stamires D. Biomass Catalytic Cracking Catalyst and Process of Preparing the Same. US Patent 9,333,494, May 10, 2016.

[115] Howell P. Preparation of Crystalline Zeolites. US Patent 3,390,958, July 2, 1968.

[116] Yildiz G, Ronsse F, Venderbosch R, van Duren R, Kersten S, Prins W. Effect of Biomass Ash in Catalytic Fast Pyrolysis of Pine Wood. Appl Catal B 2015;168-169:203-211. https://doi.org/10.1016/j.apcatb.2014.12.044

[117] Yildiz G, Ronsse F, van Duren R, Prins W. Challenges in the Design and Operation of Processes for Catalytic Fast Pyrolysis of Woody Biomass. Renewable Sustainable Energy Rev 2016;57:1596-1610. https://doi.org/10.1016/j.rser.2015.12.202

[118] Kristina L, French R, Orton K, Yung M, Johnson D, ten Dam J, Watson M, Nimlos M. In Situ and Ex Situ Catalytic Pyrolysis of Pine in a Bench-Scale Fluidized Bed Reactor System. Energy Fuels 2016;30(3):2144-2157. https://doi.org/10.1021/acs.energyfuels.5b02165

[119] Ramirez Corredores M M, Sorrells J. Fungible Bio-oil. US Patent 20130174476, July 11, 2013.

[120] Sanchez V, Engelman R, Lewis C, Moore B, Smith E. Process for Upgrading Biomass Derived Products using Liquid-Liquid Extraction. US Patent 20160317947, Nov. 3, 2016.

[121] Ritter S. Biofuel Bonanza. Chem Eng News 2007;85(26):15-24. http://pubsapp.acs.org/cen/coverstory/85/8526cover.html

[122] Loescher M. The Path to Commercialization of Drop-in Cellulosic Transportation Fuels. Proceedings of the International Conference on Thermochemical Biomass Conversion Science (tcbiomass 2013), Chicago, September 2013.

[123] Mendes F, Pinho A, Figueiredo M. Evaluation of the Impact of Temperature and Type of Catalyst on the Bio-Oil Quality Obtained by Biomass Catalytic Pyrolysis. Defect Diffus Forum 2013;334-335:13-18. https://doi.org/10.4028/www.scientific.net/DDF.334-335.13

[124] Stahl L. The Cleantech Crash. 60 Minutes, Jan. 5, 2014, https://www.cbsnews.com/news/cleantech-crash-60-minutes/ [accessed March 1, 2023].

[125] Khosla V. Open Letter to 60 Minutes and CBS, January 14, 2014, https://www.khoslaventures.com/open-letter-to-60-minutes-and-cbs [accessed March 1, 2023].

[126] Cook J. Venture Capitalist Vinod Khosla Rips 60 Minutes, Says 'Cleantech Crash' Story Uses 'Benghazi-Style Reporting', January 14, 2014, GeekWire https://www.geekwire.com/2014/ventur e-capitalist-vinod-khosla-rips-60-minutes-says-cleantech-cr ash-story-uses-benghazi-style-reporting/ [accessed March 1, 2023].

[127] Mufson, Steven, Billionaire Vinod Khosla's Big Dreams for Biofuels Fail to Catch Fire, Washington Post November 28, 2014, https://www. washingtonpost.com/business/economy/billionaire-vinod-khoslas-bi g-dreams-for-biofuels-fail-to-catch-fire/2014/11/27/04899d12-69d 7-11e4-9fb4-a622dae742a2_story.html [accessed March 1, 2023].

[128] KiOR Earnings Call Transcripts, Seeking Alpha, https://seekingalpha. com/symbol/KIOR/earnings/transcripts. [accessed March 1, 2023].

[129] Bomgardner M. KiOR in Desperate Financial Straits. Chem Eng News, 2014;92(12):16. https://cen.acs.org/articles/92/i12/ KiOR-Desperate-Financial-Straits.html

[130] Brown T. The Dangers of Relying on Next-Gen Biofuels Cost Estimates. Seeking Alpha, Nov. 8, 2013, https://seekingalpha.com/articl e/1822002-the-dangers-of-relying-on-next-gen-biofuel-cost-estimates [accessed March 1, 2023].

[131] Chatsko M. KiOR Goes Double or Nothing on Platform. Motley Fool, Sept. 27, 2013, https://www.fool.com/investing/general/2013/09/27/ kior-goes-double-or-nothing-on-platform.aspx [accessed March 1, 2023].

[132] Vasalos I, Lappas A, Kopalidou E, Kalogiannis K. Biomass Catalytic Pyrolysis: Process Design and Economic Analysis. WIREs Energy Environ 2016;5(3):370-383. https://doi.org/10.1002/wene.192

[133] Erickson P, van Asselt H, Koplow D, Lazarus M, Newell P, Oreskes N, Supran G. Why Fossil Fuel Producer Subsidies Matter. Nature 2020;578:E1-E4. http://doi.org/10.1038/s41586-019-1920-x

[134] Jewell J, Emmerling J, Vinichenko V, Bertram C, Berger L, Daly H, Keppo I, Krey V, Gernaat D, Fragkiadakis K, McCollum D, Paroussas L, Riahi K, Tavoni M, van Vuuren D. Reply to: Why Fossil Fuel Producer Subsidies Matter. Nature 2020;578:E5-E7. http://doi.org/10.1038/s41586-019-1921-9

[135] KiOR to Double Production Capacity at Mississippi Plant. Biomass Magazine, Sept. 26, 2013, http://biomassmagazine.com/articles/9482/kior-to-double-production-capacity-at-mississippi-plant [accessed March 1, 2023].

[136] O'Connor P. Optimized Catalyst for Biomass Pyrolysis. International Patent WO 2013098195, July 4, 2013; Priority Application US 2011-61580678, Dec. 28, 2011; Patent Application US 20140309467, Oct. 16, 2014.

[137] Klein T. Paul O'Connor: Don't Give Up on Advanced Biofuels. Future Fuel Strategies, Jan. 24, 2017, http://futurefuelstrategies.com/2017/01/25/paul-oconnor-dont-give-advanced-biofuels [accessed March 1, 2023].

[138] U.S. District Court for the District of Delaware October 19, 2020. Case Case No. 14-12514 (CSS), https://www.deb.uscourts.gov/sites/default/files/opinions/chief-judge-christopher-s-sontchi/kior-op-motion-enforce-plan.pdf [accessed March 1, 2023].

[139] Natesh R. KiOR—Biofuel Company Burns. Digital Initiative: Technology & Operations Management. Harvard Business School, Dec. 7, 2015, https://rctom.hbs.org/submission/kior-biofuel-company-burns [accessed March 1, 2023].

[140] Nanda R, Stuart T. KiOR: Catalyzing Clean Energy. Harvard Business School Case 809-092, March 2009 (Revised July 2009), https://www.hbs.edu/faculty/Pages/item.aspx?num=37036 [accessed March 1, 2023].

[141] Neil S. Three Critical Lessons We Can Learn from Failed Startups. Huffington Post, June 28, 2016, https://www.huffingtonpost.com/shane-paul-neil/3-critical-lessons-we-can_b_10701712.html [accessed March 1, 2023].

[142] Lane J. KIOR—The August 31, 2014, O'Connor Resignation Letter. Biofuels Digest, Sept. 9, 2014, http://www.biofuelsdigest.com/bdigest/2014/09/09/kior-the-august-31-2014-oconnor-resignation-letter/ [accessed March 1, 2023].

[143] White J. Repurposing Garbage, Chem Eng News, 2019;97(48):3. https://cen.acs.org/business/biobased-chemicals/Reactions/97/i48

[144] Ensyn, Arbec and Rémabec Begin Construction of the Cote Nord Biocrude Production Facility in Quebec. Ensyn Press Release, July 13, 2016. http://www.ensyn.com/latest-news/july-13th-2016 [accessed December 11, 2022].

[145] Freel B A, Graham R G. Apparatus for a Circulating Bed Transport Fast Pyrolysis Reactor System. US Patent 5,961,786, Oct. 5, 1999.

[146] Liew FE, Nogle R, Abdalla T, Rasor BJ, Canter C, Jensen RO, Wang L, Strutz J, Chirania P, De Tissera S, Mueller AP, Ruan Z, Gao A, Tran L, Engle NL, Bromley JC, Daniell J, Conrado R, Tschaplinski TJ, Giannone RJ, Hettich RL, Karim AS, Simpson SD, Brown SD, Leang C, Jewett MC, Köpke M. Carbon-Negative Production of Acetone and Isopropanol by Gas Fermentation at Industrial Pilot Scale. Nat Biotechnol 2022;40:335–344. https://doi.org/10.1038/s41587-021-01195-w.

[147] Service, R F. Can Biofuels Really Fly? Science 2022;376:1394-1397.

[148] Fehrenbacher K. A Biofuel Dream Gone Bad. Fortune, Dec. 4/Dec. 15, 2015, https://fortune.com/longform/kior-vinod-khosla-clean-techhttp://fortune.com/kior-vinod-khosla-clean-tech [accessed March 1, 2023].

[149] Subramaniam B, Allen D, Hii KK, Colberg J, Pradeep T. Lab to Market: Where the Rubber Meets the Road for Sustainable Chemical Technologies. ACS Sustainable Chem. Eng. 2021;9(8): 2987–2989. https://doi.org/10.1021/acssuschemeng.1c00980.

[150] Guigo N, Jérome F, Sousa AF. Biobased Furanic Derivatives for Sustainable Development. Green Chem. 2021;23:9721-9722. https://doi.org/10.1039/D1GC90124A.

[151] Chang J-N,,Qi L, Yan Y, Shi, J-W, Zhou J, Lu, M, Zhang M, Ding H-M, Chen Y, Li, S-L, Lan Y-Q. Covalent-Bonding Oxidation Group and Titanium Cluster to Synthesize a Porous Crystalline Catalyst for

Selective Photo-Oxidation Biomass Valorization. Angew. Chem. Int. Ed. 2022;61:e202209289 https://doi.org/10.1002/anie.202209289.

[152] Elliott D, Meier D, Oasmaa A, van de Beld B, Bridgwater A, Marklund M. Results of the International Energy Agency Round Robin on Fast Pyrolysis Bio-oil Production. Energy Fuels 2017;31(5):5111-5119. https://doi.org/10.1021/acs.energyfuels.6b03502

[153] Luo X, Li Y, Gupta N, Sels B, Ralph J, Shuai L. Protection Strategies Enable Selective Conversion of Biomass. Angew Chem Int Ed 2020;59(29):11704-11716. https://doi.org/10.1002/anie.201914703

[154] Mascal M. Across the Board: Mark Mascal on the Challenges of Lignin Biorefining. ChemSusChem 2020;13(1):274-277. https://doi.org/10.1002/cssc.201903042

[155] Sun Z, Cheng J, Wang D, Yuan T, Song G, Barta K. Downstream Processing Strategies for Lignin-First Biorefinery. ChemSusChem 2020;13(19):5199-5212. https://doi.org/10.1002/cssc.202001085

[156] Sun R-C, Samec JSM, Ragauskas AJ. Preface to Special Issue of ChemSusChem on Lignin Valorization: From Theory to Practice. ChemSusChem 2020;13(17):4175-4180. https://doi.org/10.1002/cssc.202001755

[157] Subbotina E, Rukkijakan, T, Marquez-Medina MD, Yu X, Johnsson M, Samec JSM. Oxidative Cleavage of C–C Bonds in Lignin. Nat. Chem. 2021 https://doi.org/10.1038/s41557-021-00783-2.

[158] Zhao D, Wang X, Miller J, Huber G. The Chemistry and Kinetics of Polyethylene Pyrolysis: A Feedstock to Produce Fuels and Chemicals. ChemSusChem 2020;13(7):1764-1774. https://doi.org/10.1002/cssc.201903434

[159] Mark LO, Cendejas MC, Hermans I. The Use of Heterogeneous Catalysis in the Chemical Valorization of Plastic Waste. ChemSusChem 2020;13(22):5808-5836. https://doi.org/10.1002/cssc.202002328

[160] Kemfert C, Präger F, Braunger I, Hoffart F M, Brauers H. The Expansion of Natural Gas Infrastructure Puts Energy Transitions at Risk. Nat. Energy 2022;7:582–587

[161] Bullard N. Clean Energy Sets $1.1 Trillion Record That's Bound to Be Broken. Bloomberg News, January 26, 2023, https://www.bloomberg.com/news/articles/2023-01-26/clean-energy-fossil-fuel-investment-tied-for-first-time-in-2022 [accessed March 1, 2023].

[162] Ostheimer G, Faulkner D. Bioenergy: The Great Continuum. Biofuels Digest, June 1, 2020, https://www.biofuelsdigest.com/bdigest/2020/06/01/bioenergy-the-great-continuum/. [accessed March 1, 2023].

[163] Bioeconomy Research & Development Act of 2020, https://www.congress.gov/bill/116th-congress/senate-bill/3734. [accessed March 1, 2023].

[164] Sustainable Chemistry Research & Development Act/National Defense Authorization Act for Fiscal Year 2021, https://www.govtrack.us/congress/bills/116/hr6395/text. [accessed March 1, 2023].

[165] Executive Order on Advancing Biotechnology and Biomanufacturing Innovation for a Sustainable, Safe, and Secure American Bioeconomy, https://www.whitehouse.gov/briefing-room/presidential-actions/2022/09/12/executive-order-on-advancing-biotechnology-and-biomanufacturing-innovation-for-a-sustainable-safe-and-secure-american-bioeconomy/. [accessed March 1, 2023].

[166] Inflation Reduction Act of 2022. https://www.democrats.senate.gov/imo/media/doc/inflation_reduction_act_of_2022.pdf. [accessed March 1, 2023].

[167] Gates B. My New Climate Book is Finally Here. Gates Notes, https://www.gatesnotes.com/Energy/My-new-climate-book-is-finally-here [accessed March 1, 2023].

[168] Danner JC, Ranon U, Stamires DN. Hyperfine, Superhyperfine, and Quadrupole Interactions of Gd3+ in YPO4 Crystals. Physical Review B 1971;3:2141-2149.

[169] Ranon U, Stamires DN. Electron Spin Resonance of Mn2+ in CdF2. Chemical Physics Letters 1968;2:286-288.

[170] Stamires DN. Electron Spin Resonance of Atomic Hydrogen Adsorbed on Porous Crystals and Perturbation of its Electronic Wave Function by

the Electric Field of the Solid Surface. Bulletin of the American Physical Society. 1965;10:700.

[171] Electron Spin Resonance Spectra of Magnetic Centers in Porous Crystals Produced by Ionizing Radiation. In Molecular Sieves, Society of Chemical Industry, London, 1968.

[172] Ranon U, Stamires DN. Determination of the Nuclear Quadrupole Moment Ratio of Gd155/Gd157 by Electron Paramagnetic Resonance. Chemical Physics Letters 1970;5:221-225.

[173] Byrd N, Kleist ST, Stamires DN. Electric and Magnetic Properties of Polymeric Organic Semiconductors. Journal of Polymer Science: Part A2 1972;10, 957-959.

[174] Ranon U, Stamires N. Crystal Field Strain Induced Effects on the EPR Spectra of Dy+3 in YVO4 Single Crystals. Bulletin of the American Physical Society. 1972;17:49.

ABOUT THE AUTHORS

Dennis Stamires (born in Greece, 1932) is an American scientist and expert in heterogeneous catalysis, solid-gas interface interactions, nuclear-electron interactions, and electron-transfer reactions. He received a B.Sc. in chemistry from the University of Leeds in 1953, an M.S. in physical chemistry at Canisius College in 1958, and a Ph.D. in chemical physics from Princeton University in 1962 working in the research group of John Turkevitch. Stamires' subsequent career was influenced by the university's efforts at that time to bring closer the disciplines and research projects of chemistry, physics, and the Princeton Institute of Advanced Studies, resulting in the formation of chemical physics as a discipline, as spearheaded by Turkevitch's advisor, Hugh Taylor, before Taylor left Princeton for Harvard University where he made similar contributions, including research collaborations between Harvard and Princeton. Stamires participated in a project on refining infrared reflectance spectroscopy of crystalline surfaces under the direction of George B. Kistiakowsky (Heterogeneous Catalysis: Selected American Histories https://pubs.acs.org/isbn/9780841207783). Among his resulting achievements, Stamires received an invitation by the Soviet Academy of Sciences to participate in an international exchange program at the Moscow Institute of Chemical Physics in collaboration with Vladimir B. Kazansky to work on the physicochemical and catalytic properties of acidic and basic porous solids. Stamires subsequently was appointed by the U.S. National Academy of Sciences to host Kazansky in a series of meetings with catalysis research groups at the University of California, Berkeley, and Stanford University.

Dr. Stamires began his independent career at Union Carbide's Linde Division, working on the electrical properties of metal-ion-exchanged synthetic faujasite-type zeolites (molecular sieves) and preparation of zeolitic solid-state electronic devices, including humidity sensors and dry-cell high-temperature batteries. Key to this work was showing experimentally that X-type wide-pore synthetic faujasite zeolites are relatively unstable and lose their crystallinity when exposed to thermal or hydrothermal treatments, a problem solved by replacing zeolite X with its isomorph zeolite Y containing a higher silica-to-alumina molar ratio. This work led to the discovery with his colleagues of the unique catalytic properties of metal-ion-exchanged Y and ultrastable or decationized Y faujasite zeolites for converting petroleum to transportation fuels and specialty chemicals that are now used in commercial fluid catalytic cracking (FCC) and hydrocracking catalysts. These substantive advances in commercial refining catalysts helped increase the volume and availability of petroleum-derived gasoline, diesel fuel, and jet fuel globally at lower costs. For these innovative scientific contributions, Stamires, Jule Rabo, and coworkers were considered by the National Academy of Sciences to be proposed as candidates for a Nobel Prize in Chemistry.

In 1965, Dr. Stamires joined the new Douglas Advanced Research Laboratory (DARL, formed by Douglas Aircraft Corp., later McDonnell Douglass and then Boeing Co.) in Huntington Beach, California. Among other projects, he worked with Nobel Laureate Willard F. Libby, recognized for his earlier work on radiocarbon dating but at the time was a passionate advocate about cleaning up air pollution in Southern California; Libby was a professor of chemistry and environmental science at the University of California, Los Angeles, and a member of the Douglas Aircraft Co. Board of Directors. Stamires worked on experimental assessment of automobile and supersonic aircraft engine exhaust gases to understand their detrimental pollution effects on stratospheric ozone using electron spin resonance (ESR) spectroscopy. These studies aided the U.S. Department of Transportation and Federal Aviation Administration decision to not allow SST aircraft to fly over the continental U.S. Stamires also worked on the construction of a high-resolution Electron Nuclear Double Resonance (ENDOR) spectrometer for improving the accuracy and resolution of regular ESR spectrometry for examining electron-nuclear interactions in single crystals, polycrystalline materials, and gaseous mixtures. The ENDOR section of the whole assembly included a low-temperature double-wall cryogenic liquid helium storage dewar located at the ESR cavity capable of cooling samples to 4.2 K. The system was designed in collaboration with August Maki at the University of California, Davis, and was also used to analyze single crystals of solid-state lasers grown at Bell Laboratories in a collaboration project with DARL. Some of the results have been published [168-174].

During the early 1960s, four prominent research projects in the fields of atmospheric chemistry and environmental science took place in California regarding innovative atmospheric decarbonization that produced effective means to help protect the planet. Notably, three of them were under the auspices of Libby to different extents. These included Libby's own environmental research group at UCLA developing the catalytic converter used to treat the exhaust gases of internal combustion automotive engines, Stamires' research at DARL identifying the reaction and destruction of ozone by the exhaust gases of supersonic (SST) aircraft that used ESR spectroscopic gas-phase techniques, F. Sherwood Rowland and Mario Molina's pioneering atmospheric chemistry research in identifying the destructive reaction pathways of ozone caused by anthropogenic chlorofluorocarbon (CFC) emissions to the atmosphere, and Stamires's confirmation of Rowland and Molina's experimental results of ozone destruction using a different ESR identification and quantification measuring system. Stamires's further work on developing a more efficient and less costly pressure swing adsorption process using ion-exchanged zeolites as sorbents provided a means to separate mercaptan and carbon dioxide from methane in the leaking gases of mature landfills. This allowed purified methane to be piped to a local natural gas supply company, turning a waste material to a commodity product. This development also further prevented unwanted pollutants from entering the atmosphere to impact global warming and to destroy ozone and thereby weaken the atmosphere's UV radiation screening effect, which would allow more radiation to reach planet surface and increase the incidence of skin cancer.

Stamires joined Filtrol/Kaiser Aluminum & Chemicals Corp. in Los Angeles in 1972, becoming Filtrol's vice president of R&D in 1979 overseeing development and production of low-cost synthetic faujasite-type zeolites and FCC specialty catalysts for producing low-sulfur and low-nitrogen fuels, including high-octane gasoline. In 1982, Stamires was offered a consulting position at AkzoNobel in the Netherlands to assist the catalyst division in reviving its struggling global business, leading to development of next-generation FCC and hydroprocessing catalysts and their production facilities. Subsequently, AkzoNobel bought Filtrol in 1989 and then sold its catalyst business to Albemarle in 2004, where Stamires remained as a full-time consultant until 2006 working on petroleum refining catalysts and new fire-retardant products.

As an environmental advocate motivated by a desire to develop cleaner burning fuels and reduce air pollution, work that was not targeted at Albemarle, Stamires joined new start-up company BIOeCON in the Netherlands as a consultant with the purpose of developing catalytic cracking processes for converting waste biomass to transportation fuels. When BIOeCON collaborated with Khosla Ventures for financing in 2007, Stamires joined the new Houston-based KiOR as a consultant with the title of Senior Fellow-Scientist and as a member of the KiOR management team, remaining in those roles until late 2013.

Throughout his career, Dr. Stamires has collaborated with global leading scientists and engineers in developing catalysts and specialty chemicals for a range of applications with societal benefits. For example, in collaboration with William Jones at the University of Cambridge he participated in research projects that led to the development of cellular drug-delivery nanocarrier materials used in chemotherapy for cancer treatments.

While Stamires was in Athens for the 1972 Posidonia International Conference and Exhibition he met with members of the science management team of the Nuclear Research Center "Demokritos," now known as the National Centre of Scientific Research "Demokritos," and proposed R&D projects for using nuclear power to develop low-cost commodity products and their subsequent commercialization to help improve Greece's economic status. Stamires' further science and technology contributions include a research effort in 1972 with archaeologist and president of the Greek National Museum, Spyridon Marinatos, to use new noninvasive acoustical holographic technology for excavations at the prehistoric city of Akrotiri on the island of Santorini. Acoustical holography was developed in the late 1960s by Alexander F. Metherell, a colleague of Stamires at DARL at that time. The main advantage of this technique was that individual treasures could be identified and precisely located underground before excavation to avoid damaging the artifacts.

In the early 1980s, Stamires visited the Greek Islands of Milos and Kimolos and observed mining of bentonite clay and worked with management and research teams in developing new commercial applications and premium products to improve the companies' clay businesses. Stamires further helped in developing bauxite ore resources in the Mt. Parnassus area for specialty aluminum products, for example in the preparation of heterogeneous catalysts and sorbents. These science and technology achievements have been a challenge in Greece owing to government rules and regulations and inadequate oversight, as described by Iacovos Vasalos in "Adventures for Creating Research Infrastructure in Greece" (ISBN: 978-960-04-5090-3).

Stamires' work is described in 630 patents and patent applications, 126 research publications, and 94 public scientific presentations. He is a member of the New York Academy of Science and member of The Circle of Hellenic Academics in Boston, and he has been a member of the American Physical Society and the American Chemical Society. Stamires during his long scientific career has received many awards as well as commendations for his candor in sticking with the facts and his fairness to coworkers. Stamires continues working as a passionate advocate for protecting the environment (land, water, and air) to avoid catastrophic climate change and extinction of life.

Stephen K. Ritter (born in North Carolina, USA, in 1963) received a B.S. in industrial chemistry (1986), B.A. in technical writing and editing (1989), and M.S. in nuclear chemistry with a focus on radon assessment (1990), all from Western Carolina University. Following military service in the U.S. Marine Corps, he received a Ph.D. degree in inorganic chemistry at Wake Forest University in 1993 and conducted postdoctoral research at the University of Idaho, applying main-group fluorine chemistry to the development of new inorganic and organometallic polymeric materials. Dr. Ritter started his independent career at the American Chemical Society in 1994 as an assistant editor at *Chemical & Engineering News*. Over the years his roles evolved as an expert on topics of inorganic chemistry, energy, and environmental science, with broad coverage of green chemistry, biomass conversion, natural resource management, and sustainability science, rising to become Senior Editor with more than 1,400 published articles to his credit. In January 2018, Dr. Ritter joined ACS Global Journals Development as Managing Editor for core inorganic and organic journals.

Printed in the United States
by Baker & Taylor Publisher Services